Probability Theory
and Combinatorial
Optimization

CBMS-NSF REGIONAL CONFERENCE SERIES IN APPLIED MATHEMATICS

A series of lectures on topics of current research interest in applied mathematics under the direction of the Conference Board of the Mathematical Sciences, supported by the National Science Foundation and published by SIAM.

GARRETT BIRKHOFF, *The Numerical Solution of Elliptic Equations*
D. V. LINDLEY, *Bayesian Statistics, A Review*
R. S. VARGA, *Functional Analysis and Approximation Theory in Numerical Analysis*
R. R. BAHADUR, *Some Limit Theorems in Statistics*
PATRICK BILLINGSLEY, *Weak Convergence of Measures: Applications in Probability*
J. L. LIONS, *Some Aspects of the Optimal Control of Distributed Parameter Systems*
ROGER PENROSE, *Techniques of Differential Topology in Relativity*
HERMAN CHERNOFF, *Sequential Analysis and Optimal Design*
J. DURBIN, *Distribution Theory for Tests Based on the Sample Distribution Function*
SOL I. RUBINOW, *Mathematical Problems in the Biological Sciences*
P. D. LAX, *Hyperbolic Systems of Conservation Laws and the Mathematical Theory of Shock Waves*
I. J. SCHOENBERG, *Cardinal Spline Interpolation*
IVAN SINGER, *The Theory of Best Approximation and Functional Analysis*
WERNER C. RHEINBOLDT, *Methods of Solving Systems of Nonlinear Equations*
HANS F. WEINBERGER, *Variational Methods for Eigenvalue Approximation*
R. TYRRELL ROCKAFELLAR, *Conjugate Duality and Optimization*
SIR JAMES LIGHTHILL, *Mathematical Biofluiddynamics*
GERARD SALTON, *Theory of Indexing*
CATHLEEN S. MORAWETZ, *Notes on Time Decay and Scattering for Some Hyperbolic Problems*
F. HOPPENSTEADT, *Mathematical Theories of Populations: Demographics, Genetics and Epidemics*
RICHARD ASKEY, *Orthogonal Polynomials and Special Functions*
L. E. PAYNE, *Improperly Posed Problems in Partial Differential Equations*
S. ROSEN, *Lectures on the Measurement and Evaluation of the Performance of Computing Systems*
HERBERT B. KELLER, *Numerical Solution of Two Point Boundary Value Problems*
J. P. LASALLE, *The Stability of Dynamical Systems*
D. GOTTLIEB AND S. A. ORSZAG, *Numerical Analysis of Spectral Methods: Theory and Applications*
PETER J. HUBER, *Robust Statistical Procedures*
HERBERT SOLOMON, *Geometric Probability*
FRED S. ROBERTS, *Graph Theory and Its Applications to Problems of Society*
JURIS HARTMANIS, *Feasible Computations and Provable Complexity Properties*
ZOHAR MANNA, *Lectures on the Logic of Computer Programming*
ELLIS L. JOHNSON, *Integer Programming: Facets, Subadditivity, and Duality for Group and Semi-Group Problems*
SHMUEL WINOGRAD, *Arithmetic Complexity of Computations*
J. F. C. KINGMAN, *Mathematics of Genetic Diversity*
MORTON E. GURTIN, *Topics in Finite Elasticity*
THOMAS G. KURTZ, *Approximation of Population Processes*
JERROLD E. MARSDEN, *Lectures on Geometric Methods in Mathematical Physics*
BRADLEY EFRON, *The Jackknife, the Bootstrap, and Other Resampling Plans*
M. WOODROOFE, *Nonlinear Renewal Theory in Sequential Analysis*
D. H. SATTINGER, *Branching in the Presence of Symmetry*
R. TEMAM, *Navier–Stokes Equations and Nonlinear Functional Analysis*

MIKLÓS CSÖRGO, *Quantile Processes with Statistical Applications*
J. D. BUCKMASTER AND G. S. S. LUDFORD, *Lectures on Mathematical Combustion*
R. E. TARJAN, *Data Structures and Network Algorithms*
PAUL WALTMAN, *Competition Models in Population Biology*
S. R. S. VARADHAN, *Large Deviations and Applications*
KIYOSI ITÔ, *Foundations of Stochastic Differential Equations in Infinite Dimensional Spaces*
ALAN C. NEWELL, *Solitons in Mathematics and Physics*
PRANAB KUMAR SEN, *Theory and Applications of Sequential Nonparametrics*
LÁSZLÓ LOVÁSZ, *An Algorithmic Theory of Numbers, Graphs and Convexity*
E. W. CHENEY, *Multivariate Approximation Theory: Selected Topics*
JOEL SPENCER, *Ten Lectures on the Probabilistic Method*
PAUL C. FIFE, *Dynamics of Internal Layers and Diffusive Interfaces*
CHARLES K. CHUI, *Multivariate Splines*
HERBERT S. WILF, *Combinatorial Algorithms: An Update*
HENRY C. TUCKWELL, *Stochastic Processes in the Neurosciences*
FRANK H. CLARKE, *Methods of Dynamic and Nonsmooth Optimization*
ROBERT B. GARDNER, *The Method of Equivalence and Its Applications*
GRACE WAHBA, *Spline Models for Observational Data*
RICHARD S. VARGA, *Scientific Computation on Mathematical Problems and Conjectures*
INGRID DAUBECHIES, *Ten Lectures on Wavelets*
STEPHEN F. MCCORMICK, *Multilevel Projection Methods for Partial Differential Equations*
HARALD NIEDERREITER, *Random Number Generation and Quasi-Monte Carlo Methods*
JOEL SPENCER, *Ten Lectures on the Probabilistic Method, Second Edition*
CHARLES A. MICCHELLI, *Mathematical Aspects of Geometric Modeling*
ROGER TEMAM, *Navier–Stokes Equations and Nonlinear Functional Analysis, Second Edition*
GLENN SHAFER, *Probabilistic Expert Systems*
PETER J. HUBER, *Robust Statistical Procedures, Second Edition*
J. MICHAEL STEELE, *Probability Theory and Combinatorial Optimization*
WERNER C. RHEINBOLDT, *Methods for Solving Systems of Nonlinear Equations, Second Edition*
J. M. CUSHING, *An Introduction to Structured Population Dynamics*
TAI-PING LIU, *Hyperbolic and Viscous Conservation Laws*
MICHAEL RENARDY, *Mathematical Analysis of Viscoelastic Flows*
GÉRARD CORNUÉJOLS, *Combinatorial Optimization: Packing and Covering*
IRENA LASIECKA, *Mathematical Control Theory of Coupled PDEs*
J. K. SHAW, *Mathematical Principles of Optical Fiber Communications*
ZHANGXIN CHEN, *Reservoir Simulation: Mathematical Techniques in Oil Recovery*
ATHANASSIOS S. FOKAS, *A Unified Approach to Boundary Value Problems*
MARGARET CHENEY AND BRETT BORDEN, *Fundamentals of Radar Imaging*
FIORALBA CAKONI, DAVID COLTON, AND PETER MONK, *The Linear Sampling Method in Inverse Electromagnetic Scattering*
ADRIAN CONSTANTIN, *Nonlinear Water Waves with Applications to Wave-Current Interactions and Tsunamis*
WEI-MING NI, *The Mathematics of Diffusion*
ARNULF JENTZEN AND PETER E. KLOEDEN, *Taylor Approximations for Stochastic Partial Differential Equations*
FRED BRAUER AND CARLOS CASTILLO-CHAVEZ, *Mathematical Models for Communicable Diseases*
PETER KUCHMENT, *The Radon Transform and Medical Imaging*

J. Michael Steele
Wharton School
University of Pennsylvania
Philadelphia, Pennsylvania

Probability Theory and Combinatorial Optimization

siam.

SOCIETY FOR INDUSTRIAL AND APPLIED MATHEMATICS
PHILADELPHIA

Copyright ©1997 by the Society for Industrial and Applied Mathematics.

10 9 8 7 6 5

All rights reserved. Printed in the United States of America. No part of this book may be reproduced, stored, or transmitted in any manner without the written permission of the publisher. For information, write to the Society for Industrial and Applied Mathematics, 3600 Market Street, 6th Floor, Philadelphia, PA 19104-2688 USA.

Library of Congress Cataloging-in-Publication Data

Steele, J. Michael.
    Probability theory and combinatorial optimization / J. Michael Steele.
      p. cm. — (CBMS-NSF regional conference series in applied
      mathematics ; 69)
    Includes bibliographical references and index.
    ISBN 978-0-898713-80-0 (pbk.)
      1. Probabilities. 2. Combinatorial optimization. I. Title. II. Series.

QA273.45.S74   1996
519.7—dc20

                                                      96-42685

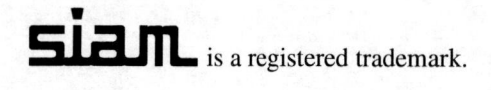

siam is a registered trademark.

Contents

Preface   ix

Chapter 1. First View of Problems and Methods   1
    1.1. A first example: Long common subsequences.   1
    1.2. Subadditivity and expected values.   2
    1.3. Azuma's inequality and a first application.   4
    1.4. A second example: The increasing-subsequence problem.   6
    1.5. Flipping Azuma's inequality.   10
    1.6. Concentration on rates.   13
    1.7. Dynamic programming.   16
    1.8. Kingman's subadditive ergodic theorem.   18
    1.9. Observations on subadditive subsequences.   21
    1.10. Additional notes.   24

Chapter 2. Concentration of Measure and the Classical Theorems   27
    2.1. The TSP and a quick application of Azuma's inequality.   27
    2.2. Easy size bounds.   30
    2.3. Another mean Poissonization.   30
    2.4. The Beardwood–Halton–Hammersley theorem.   33
    2.5. Karp's partitioning algorithms.   40
    2.6. Introduction to the space-filling curve heuristic.   41
    2.7. Asymptotics for the space-filling curve heuristic.   43
    2.8. Additional notes.   49

Chapter 3. More General Methods   53
    3.1. Subadditive Euclidean functionals.   53
    3.2. Examples: Good, bad, and forthcoming.   58
    3.3. A general L^∞ bound.   59
    3.4. Simple subadditivity and geometric subadditivity.   60

CONTENTS

- 3.5. A concentration inequality. 61
- 3.6. Minimal matching. 63
- 3.7. Two-sided bounds and first consequences. 65
- 3.8. Rooted duals and their applications. 68
- 3.9. Lower bounds and best possibilities. 71
- 3.10. Additional remarks. 75

Chapter 4. Probability in Greedy Algorithms and Linear Programming   77

- 4.1. Assignment problem. 77
- 4.2. Simplex method for theoreticians. 82
- 4.3. Dyer–Frieze–McDiarmid inequality. 84
- 4.4. Dealing with integral constraints. 88
- 4.5. Distributional bounds. 89
- 4.6. Back to the future. 91
- 4.7. Additional remarks. 93

Chapter 5. Distributional Techniques and the Objective Method   95

- 5.1. Motivation for a method. 95
- 5.2. Searching for a candidate object. 96
- 5.3. Topology for nice sets. 100
- 5.4. Information on the infinite tree. 102
- 5.5. Dénoument. 103
- 5.6. Central limit theory. 108
- 5.7. Conditioning method for independence. 110
- 5.8. Dependency graphs and the CLT. 112
- 5.9. Additional remarks. 117

Chapter 6. Talagrand's Isoperimetric Theory   119

- 6.1. Talagrand's isoperimetric theory. 119
- 6.2. Two geometric applications of the isoperimetric inequality. ... 121
- 6.3. Application to the longest-increasing-subsequence problem. ... 125
- 6.4. Proof of the isoperimetric inequality. 128
- 6.5. Application and comparison in the theory of hereditary sets. ... 131
- 6.6. Suprema of linear functionals. 133
- 6.7. Tail of the assignment problem. 135
- 6.8. Further applications of Talagrand's isoperimetric inequalities. .. 140
- 6.9. Final considerations on related work. 140

Bibliography   143

Index   157

Preface

This monograph provides an introduction to the state of the art of the probability theory that is most directly applicable to combinatorial optimization. The questions that receive the most attention are those that deal with discrete optimization problems for points in Euclidean space, such as the minimum spanning tree, the traveling-salesman tour, and minimal-length matchings. Still, there are several nongeometric optimization problems that receive full treatment, and these include the problems of the longest common subsequence and the longest increasing subsequence. Throughout the monograph, our analysis focuses on understanding the behavior of the value of the objective function of the optimization problem; and, in almost all cases, we concentrate our efforts on the most elementary stochastic models for the values of the problem inputs. The philosophy that guides the exposition is that analysis of concrete problems is the most effective way to explain even the most general methods or abstract principles.

There are three fundamental probabilistic themes that are examined through our concrete investigations. First, there is a systematic exploitation of martingales. Over the last ten years, many investigators of problems of combinatorial optimization have come to count on martingale inequalities as versatile tools which let us show that many of the naturally occuring random variables of combinatorial optimization are sharply concentrated about their means—a phenomenon with numerous practical and theoretical consequences.

The second theme that is explored is the systematic use of subadditivity of several flavors, ranging from the naïve subadditivity of real sequences to the subtler subadditivity of stochastic processes. By and large, subadditivity offers only elementary tools, but on remarkably many occasions, such tools provide the key organizing principle in the attack on problems of nearly intractable difficulty.

The third and deepest theme developed here concerns the application of Talagrand's isoperimetric theory of concentration inequalities. This new theory is reshaping almost everything that is known in the probability theory of combinatorial optimization. The treatment given here deals with only a small part of Talagrand's theory, but the reader will find considerable coaching on how to use some of the most important ideas from that theory. In particular, we show how Talagrand's convex distance inequality can be used to obtain state-of-the-art concentration inequalities for the traveling-salesman problem, the assignment problem, and the longest-increasing-subsequence problem.

Many individuals have helped the framing of this monograph, and special thanks are due my former students Timothy Snyder and Jun Gao as well as the participants in the courses on this material at Princeton, Columbia, and the University of Pennsylvania. Of the many researchers who have made preprints of their work available to me for discussion in this monograph, I particularly thank David Aldous, Kenneth Alexander, Alan Frieze, Patrick Jaillet, Harry Kesten, Sungchul Lee, Matthew Penrose, Wan-Soo Rhee, Michel Talagrand, and Joseph Yukich. I also thank Anant Godbole for organizing the NSF–CBMS workshop at Michigan Technical University in the summer of 1995, for which the penultimate version of this monograph was prepared.

Finally, I am pleased to thank my wife Patricia for her kind willingness to put up with great stacks of paper all over our house and for her tolerance of my periodic hogging of the family computer.

CHAPTER 1
First View of Problems and Methods

Our path begins with the investigation of two well-studied problems of discrete optimization, the long-common-subsequence problem and the longest-increasing-subsequence problem. These problems are easily explained, and they offer a concrete basis for the illustration the remarkable effectiveness of subadditive methods and martingale inequalities. Our study of the increasing-subsequence problem also provides a vehicle for discussion of the trickier "flipping method," which can be applied in many combinatorial investigations. Finally, we give a systematic treatment of some more refined results on subadditive sequences.

1.1. A first example: Long common subsequences.

The long-common-subsequence problem considers two finite strings from a finite alphabet and asks about the size of the largest string that appears as a subsequence in each of the two strings. The problem has emerged more or less independently in several remarkably disparate areas, including the comparison of versions of computer programs, cryptographic snooping, the mathematical analysis of bird songs, and molecular biology.

The biological story behind the problem is that long molecules such as proteins and nucleic acids can be viewed schematically as sequences from a finite alphabet; in particular, DNA is commonly represented as a string from the four letter alphabet {A, C, T, G}, where each of the symbols stands for one of the DNA bases. From an evolutionary point of view, it is natural to compare the DNA of two organisms by finding their closest common ancestors, and, if one takes the restricted view that these molecules evolve only through the process of inserting new symbols in the representing strings, then the ancestors of a long molecule are just the substrings of the string that represents the molecule. Thus the length of the longest common subsequence shared by the strings X and Y is a reasonable measure of how closely X and Y are related in evolutionary terms. Naturally, this metric is subject to criticism, and it is not currently the most popular method for DNA comparison; but the metric still has appeal, and its biological basis is as sound as many of the other metrics that have been applied.

The probabilistic analysis of the problem of long common subsequences was begun in Chvátal and Sankoff (1975). They considered the situation where the letters that make up the strings are chosen by repeated independent sampling from a given alphabet. Specifically, they consider two sequences $\{X_i\}$ and $\{Y_i\}$ of independent identically distributed random variables with values in a finite alphabet, and they introduce the random variables L_n defined by

$$L_n = \max\{k : X_{i_1} = Y_{j_1}, X_{i_2} = Y_{j_2}, \ldots, X_{i_k} = Y_{j_k}\},$$

where the maximum is taken over all pairs of subsequences $1 \leq i_1 < i_2 < \cdots < i_k \leq n$ and $1 \leq j_1 < j_2 < \cdots < j_k \leq n$. Or, in other words, L_n is the largest cardinality of any subsequence common to the sequences $\{X_1, X_2, \ldots, X_n\}$ and $\{Y_1, Y_2, \ldots, Y_n\}$; see Figure 1.1. We will develop some of the theory behind this variable, and along the way we will find some of the central tools that relate probability theory and combinatorial optimization.

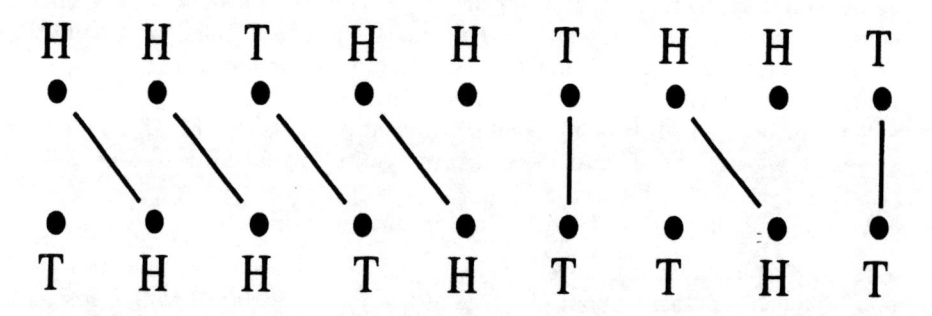

FIG. 1.1. *Illustration of the longest common subsequence for coin flips.*

1.2. Subadditivity and expected values.

Subadditivity arises in many optimization problems because they often permit decompositions into subproblems with solutions that can be recombined to give a suboptimal solution to the original problem. To see how subadditivity (or the closely related *superadditivity*) enters into the analysis of L_n, we first bundle X_i and Y_i into $Z_i = (X_i, Y_i)$ and write $L_n = L_n(Z_1, Z_2, \ldots, Z_n)$ as a function of the random variables Z_i. By the suboptimality of the subsequences formed by adjoining the longest common substring of $\{Z_1, Z_2, \ldots, Z_m\}$ and of $\{Z_{m+1}, Z_{m+2}, \ldots, Z_{m+n}\}$ to get a common substring of $\{Z_1, Z_2, \ldots, Z_{m+n}\}$, we find the natural relation

$$L_{m+n}(Z_1, Z_2, \ldots, Z_{m+n}) \geq L_m(Z_1, Z_2, \ldots, Z_m) + L_n(Z_{m+1}, Z_{m+2}, \ldots, Z_{m+n}).$$

When we let $a_n = -EL_n(Z_1, Z_2, \ldots, Z_n)$, we see that the sequence $\{a_n\}$ is subadditive. The surprising fact is that the subadditivity one obtains in this semiautomatic way makes light work of limit results that would otherwise appear to require considerable ingenuity. Perhaps the most informative fact about subadditive sequences is very simple convergence lemma.

LEMMA 1.2.1 (Fekete (1923)). *If a sequence of real numbers $\{a_n\}$ satisfies the subadditivity condition*

$$a_{m+n} \leq a_m + a_n,$$

then

$$\lim_{n \to \infty} a_n/n = \inf a_n/n.$$

Proof. We first set $\gamma = \inf a_n/n$ and consider the case where $\gamma > -\infty$. For any $\epsilon > 0$, we can find a k such that $a_k \leq (\gamma + \epsilon)k$. By subadditivity, we then have for all n and k that $a_{nk} \leq na_k$, so we also have $\gamma = \liminf a_n/n$. Generally, for any m, we can write $m = nk + j$ with $0 \leq j < k$, so we have

$$a_m = a_{nk+j} \leq a_{nk} + a_j \leq (\gamma + \epsilon)nk + \max_{0 \leq j < k} a_j.$$

Dividing by m and taking the limit superior yields

$$\limsup a_m/m \leq \gamma + \epsilon \leq \liminf a_m/m + \epsilon.$$

By the arbitrariness of $\epsilon > 0$, the $\gamma > -\infty$ case of the lemma is complete. One can use a similar argument to complete the easier case $\gamma = -\infty$.

When we apply Fekete's lemma to $a_n = -EL_n(Z_1, Z_2, \ldots, Z_n)$, we obtain our first limit result:

(1.1) $$\lim_{n \to \infty} \frac{1}{n} E(L_n) = \gamma = \sup_n \frac{1}{n} E(L_n).$$

One feature of limit results like (1.1) is the attention that they focus on the limiting constant whose existence is proclaimed. The determination of the limiting constant is often difficult. In fact, there are fewer than a handful of cases where we are able to calculate the limiting constant obtained by a subadditivity argument; even good approximations of the constants often require considerable ingenuity.

The computation of the γ of (1.1) is particularly tempting when the choices are made with equal probability from an alphabet of size k; and if c_k denotes the corresponding value of γ, there is a special charm to $k = 2$, the case of the longest common subsequence of two sequences of independent coin flips. Over the last twenty years, a good bit of energy has been invested in the determination of c_2; and, so far, all of the results are compatible with the conjecture from Steele (1982b) that $c_2 = 2/(1 + \sqrt{2}) \approx 0.828427$. The best bounds that are known with certainty are $0.7615 \leq c_2 \leq 0.8376$, due to Deken (1979) and Dančík and Paterson (1994). Simulations of Eggert and Waterman cited by Alexander (1994b) suggest that one has $0.8079 \leq c_2 \leq 0.8376$ with at least 95% confidence.

A proof of the conjecture that $c_2 = 2/(1+\sqrt{2})$ is likely to require considerable ingenuity, but to be honest, the nonnumerical evidence for the conjecture is not strong at this point. Even the numerical evidence is not all rosy; Rinsma-Melchert (1993) cites an unpublished 1988 Monte Carlo study of D. B. McKay that suggests a value of c_2 that is closer to 0.81 than to $2/(1 + \sqrt{2})$. One may be able to disprove the conjecture by pushing harder on the methods that are currently available.

There is one further problem in the theory of the long-common-subsequence problem that has not been studied so hard and yet for which there are some intriguing suggestions. The problem concerns the growth rate of the variance, and the basic issue is to decide if the variance grows linearly with n. There is simulation evidence in Waterman (1994) for this conjecture, and there is added spice to this suggestion from the fact that Chvátal and Sankoff (1975) conjectured on the basis of their simulations that the variance was of much lower order.

1.3. Azuma's inequality and a first application.

Some of the most striking applications of martingales in the theory of combinatorial optimization are based on the large-deviation inequalities due to Hoeffding (1963) and Azuma (1967). These inequalities are usually expressed in terms of independent random variables or in terms of martingale differences, but the results call more directly on a weaker property that expresses a form of multiple orthogonality. If $\{d_i : 1 \leq i \leq n\}$ are random variables such that

$$E d_{i_1} d_{i_2} \cdots d_{i_k} = 0 \quad \text{for all } 1 \leq i_1 < i_2 < \cdots < i_k \leq n,$$

then $\{d_i : 1 \leq i \leq n\}$ is called a *multiplicative* system. A key property of such a system is that for any sequences of constants $\{a_i\}$ and $\{b_i\}$, we have the identity

$$(1.2) \qquad E \prod_{i=1}^{n} \{a_i + b_i d_i\} = \prod_{i=1}^{n} a_i.$$

One well-worn path to tail bounds on sums of random variables goes through a bound on the moment-generating function, and the identity (1.2) works perfectly with this plan. We first note that the convexity of $f(x) = e^{ax}$ tells us that on the interval $[-1, 1]$, the graph of $f(x)$ lies below the line through $(-1, e^{-a})$ and $(1, e^a)$, so as illustrated in Figure 1.2, we have

$$e^{ax} \leq (x+1)(e^a - e^{-a})/2 + e^{-a} \quad \text{for } -1 \leq x \leq 1,$$

or, in a tidier form,

$$e^{ax} \leq \cosh a + x \sinh a \quad \text{for } -1 \leq x \leq 1.$$

If we now let $x = d_i/\|d_i\|_\infty$ and $a = t\|d_i\|_\infty$, we find

$$\exp\left(t \sum_{i=1}^{n} d_i\right) \leq \prod_{i=1}^{n} \left(\cosh t\|d_i\|_\infty + d_i \sinh t \frac{\|d_i\|_\infty}{\|d_i\|_\infty}\right).$$

When we take expectations and use (1.2), we find

$$(1.3) \qquad E \exp\left(t \sum_{i=1}^{n} d_i\right) \leq \prod_{i=1}^{n} \cosh(t\|d_i\|_\infty),$$

so, by the elementary bound $\cosh x \leq e^{x^2/2}$, we have

$$(1.4) \qquad E \exp\left(t \sum_{i=1}^{n} d_i\right) \leq \exp\left(\frac{1}{2} t^2 \sum_{i=1}^{n} \|d_i\|_\infty^2\right).$$

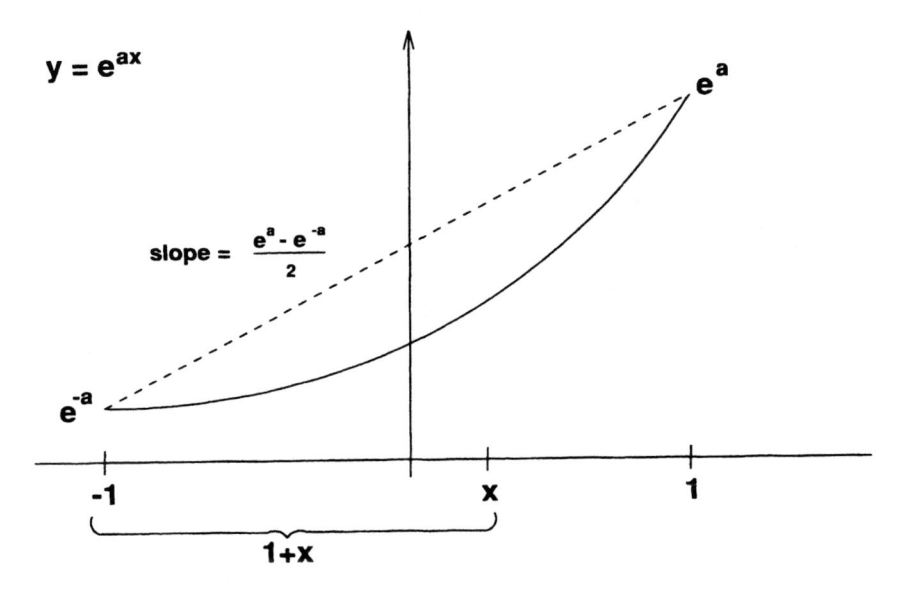

FIG. 1.2. *Geometry behind Azuma's inequality.*

Inequality (1.3) expresses the important fact that martingales with bounded differences (and similar multiplicative systems) have a moment-generating function that is bounded by the moment-generating function of a sum of scaled Bernoulli random variables, and (1.4) then exploits the fact that Bernoulli variables are sub-Gaussian. Thus the path from (1.4) to our basic tail bound is the same that one uses in the case of independent random variables. From (1.4) and Markov's inequality, we find for all $t \geq 0$ that

$$P\left(\sum_{i=1}^{n} d_i \geq \lambda\right) \leq e^{-\lambda t} \exp\left(\frac{t^2}{2}\sum_{i=1}^{n} \|d_i\|_\infty^2\right),$$

so letting $t = \lambda(\sum_{i=1}^{n} \|d_i\|_\infty^2)^{-1}$, we obtain our basic result.

THEOREM 1.3.1 (Azuma's inequality). *For random variables $\{d_i\}$ that satisfy the product identity (1.2), such as martingale differences, we have*

$$P\left(\left|\sum_{i=1}^{n} d_i\right| \geq \lambda\right) \leq 2\exp\left(-\lambda^2 / \left(2\sum_{i=1}^{n} \|d_i\|_\infty^2\right)\right).$$

To make good use of this inequality in the study of L_n, we call on an old recipe that goes back to J. Doob that shows us how to make a martingale out of any random variable. We first let $\mathcal{F}_k = \sigma\{Z_1, Z_2, \ldots, Z_k\}$, so $\mathcal{F}_k$ is the sigma-field generated by the first k of the $\{Z_i\}$'s, and then we set

(1.5) $\qquad d_i = E(L_n|\mathcal{F}_i) - E(L_n|\mathcal{F}_{i-1}).$

The sequence $\{d_i\}$ is easily checked to be a martingale-difference sequence adapted to the increasing sequence of sigma-fields $\{\mathcal{F}_i\}$, and the key relationship to the original variable is given by

$$L_n - EL_n = \sum_{i=1}^{n} d_i.$$

The real trick to exploiting this generic representation is to find some way to show that the d_i's are bounded. To do this, it is useful to introduce a new sequence of independent random variables $\{\hat{Z}_i\}$ with the same distribution as the original $\{Z_i\}$. The point of these new variables is that since $\mathcal{F}_i$ has no information about $\hat{Z}_i$, we have

$$E(L_n(Z_1, Z_2, \ldots, Z_i, \ldots, Z_n)|\mathcal{F}_{i-1}) = E(L_n(Z_1, Z_2, \ldots, Z_{i-1}, \hat{Z}_i, Z_{i+1}, \ldots, Z_n)|\mathcal{F}_i),$$

and this representation then lets us rewrite the expression for d_i in terms of a single conditional expectation:

$$d_i = E(L_n(Z_1, Z_2, \ldots, Z_i, \ldots, Z_n) - L_n(Z_1, Z_2, \ldots, Z_{i-1}, \hat{Z}_i, Z_{i+1}, \ldots, Z_n)|\mathcal{F}_i).$$

The value of the last representation should now be clear because the definition of the long-common-subsequence problem lets us check that

$$|L_n(Z_1, Z_2, \ldots, Z_i, \ldots, Z_n) - L_n(Z_1, Z_2, \ldots, Z_{i-1}, \hat{Z}_i, Z_{i+1}, \ldots, Z_n)| \leq 2.$$

Since conditional expectations cannot increase the L_∞ norm, we find that for all $1 \leq i \leq n$, we have $\|d_i\|_\infty \leq 2$. Finally, by Azuma's inequality, we have a useful tail bound for the long-common-subsequence problem:

$$P(|L_n - EL_n| \geq t) \leq 2\exp(-t^2/8n).$$

From this inequality, one can extract much of what one might want to know about the behavior of L_n, and in particular, the straightforward application of the Borel–Cantelli lemma with $t_n = \sqrt{8n(\log n)(1+\epsilon)}$ tells us that with probability one we have $L_n - EL_n = O(\sqrt{n \log n})$. Since we already know that $EL_n/n \to \gamma$, our almost sure $O(\sqrt{n \log n})$ bound is certainly good enough to tell us that $L_n/n \to \gamma$ with probability one, and it would tell us more except that the subadditivity based limit $EL_n/n \to \gamma$ does not come with any rate result. Quite recently, techniques have been developed that do give us more information on the rate at which EL_n/n converges to γ. This is a topic to which we will return shortly.

1.4. A second example: The increasing-subsequence problem.

Erdös and Szekeres (1935) showed that any sequence $\{a_j\}$ of $nm+1$ distinct real numbers must contain an increasing subsequence of length $n+1$ or a decreasing subsequence of length $m+1$. There are many proofs of this useful result, but one of the most charming is a pigeonhole argument due to Hammersley (1972).

Think of the sequence $\{a_j\}$ as a pile of numbered checkers where a_j is the number on the jth checker. We build a set of new piles as follows:

1. Let a_1 start the first new pile.

2. For each subsequent j, place a_j on top of the first pile for which a_j is larger than the checker now on top.

As one proceeds with the $nm+1$ elements of $\{a_j\}$, either a pile grows to height of at least $n+1$ or else we begin at least $m+1$ piles. The proof is completed by two observations, the first of which is just that each pile is an increasing subsequence. The second observation is that if there are $m+1$ piles, then there is a decreasing subsequence of length $m+1$ since the only reason to move to the kth pile with the jth checker is that there is a checker i that precedes j that is in the $(k-1)$th pile and for which $a_i > a_j$.

The most natural probabilistic questions in the theory of monotone subsequences concern the length of the longest increasing subsequence of n points chosen at random from the unit interval. Formally, we let X_i, $1 \leq i < \infty$, be independent random variables with the uniform distribution and consider the random variable defined by

$$I_n = \max\{k: X_{i_1} < X_{i_2} < \cdots < X_{i_k}, 1 \leq i_1 < i_2 < \cdots < i_k \leq n\}.$$

To get a lower bound on EI_n, we let I'_n be defined correspondingly as the length of the longest decreasing subsequence. By the Erdös–Szekeres theorem, we have $\max(I_n, I'_n) \geq \sqrt{n}$, so certainly $I_n + I'_n \geq \sqrt{n}$. Since $EI_n = EI'_n$ by symmetry, we have the bound $EI_n \geq \frac{1}{2}\sqrt{n}$. To get an upper bound for EI_n that is also of order $O(\sqrt{n})$ is also relatively easy, and en route we will get a useful tail bound for I_n.

LEMMA 1.4.1.
$$(1.6) \qquad P(I_n \geq 2e\sqrt{n}) < \exp(-2e\sqrt{n}).$$

Proof. First, we note that there are $\binom{n}{k}$ subsequences of length k contained in the n sequence $\{X_1, X_2, \ldots, X_n\}$ and each of these has probability $1/k!$ of being monotone increasing. By Boole's inequality, we therefore have

$$P(I_n \geq k) \leq \binom{n}{k}/k!.$$

If we let k be the least integer as great as $2e\sqrt{n}$ and apply the crude estimate $k! \geq k^k/e^k$, the lemma is completed with an easy calculation.

This exponential inequality has a striking symmetry, and it certainly suffices to show that $EI_n = O(\sqrt{n})$, but if one is mainly interested in EI_n, a somewhat sharper path combines the elementary bound

$$EI_n \leq k + nP(I_n \geq k)$$

with the fact that Stirling's formula tells us that for $k \sim c\sqrt{n}$ we have $\binom{n}{k}/k! = o((e/c)^{cn})$ to find

$$(1.7) \qquad EI_n \leq c\sqrt{n} \quad \text{for all } c > e \text{ and } n \geq n(c) \text{ sufficiently large.}$$

With both upper and lower bounds of order $\sqrt{n}$, one cannot help but suspect that the ratio $EI_n/\sqrt{n}$ has a genuine limit. Because the growth of I_n is of the order of $\sqrt{n}$, subadditivity techniques may seem inappropriate, but there is a device that brings subadditivity back into the picture. What we need is an embedding of I_n into a process that evolves at the more traditional linear rate.

Consider the Poisson point process Π on $\mathbb{R}^2$ with unit intensity. Thus if $A \subset \mathbb{R}^2$ is a measurable set, then $\Pi(A)$ is a random finite subset of $\mathbb{R}^2$, and the cardinality of $\Pi(A)$ is a random variable with the Poisson distribution with parameter $\mu = \lambda(A)$, where λ denotes Lebesgue measure on $\mathbb{R}^2$. Our problem on monotone increasing subsequences can now easily be rephrased in terms of the Poisson process.

We call a finite sequence of points $(x_1, y_1), (x_2, y_2), \ldots, (x_n, y_n)$ in $\mathbb{R}^2$ an *increasing* sequence provided $x_1 < x_2 < \cdots < x_n$ and $y_1 < y_2 < \cdots < y_n$, and we will now investigate the random variable $I(t)$ that is defined to be the cardinality of the largest increasing sequence of points in $\Pi[0,t]^2$. Before digging into the analysis of $I(t)$, we should note that I_n has the same distribution as $I(t)$ conditioned on the event $|\Pi[0,t]^2| = n$. Thus if we set $a_n = EI_n$, we have

$$(1.8) \qquad EI(t) = e^{-t^2} \sum_{k=0}^{\infty} \frac{t^{2k}}{k!} a_k.$$

Our plan is to use subadditivity to determine the asymptotics of $EI(t)$ and then exploit (1.8) to get the asymptotics of a_k.

To see why this is reasonable, we first let N be a Poisson random variable with mean t^2 and note that (1.8) just says

$$EI(t) = Ea_N.$$

Since we expect to show that $a_k \sim ck^{1/2}$, it is reasonable to suspect as well that $Ea_N \sim cEN^{1/2}$. Further, since N is a Poisson random variable with mean t^2, the expectation $EN^{1/2}$ grows linearly in t. Thus we are led to guess that $EI(t)$ grows linearly, and with this guess behind us, we see that $EI(t)$ is a reasonable candidate for analysis by subadditive methods.

We write $I(s,t)$ for the cardinality of the largest increasing sequence in $\Pi(s,t]^2$. Here we are a bit picky and consider the half-open square $(s,t]^2$. This decision will make life a little easier later, though for the moment it is unessential. The key observation illustrated by Figure 1.3 is that by suboptimality we have

$$(1.9) \qquad I(0,s) + I(s, s+t) \leq I(0, s+t).$$

Since $EI(s, s+t) = EI(0,t)$, if we write $f(t) = EI(0,t)$, then

$$f(s) + f(t) \leq f(s+t).$$

Since $f(t)$ is continuous, we find from the real-variable form of the subadditivity-limit lemma that

$$\lim_{t \to \infty} EI(t)/t = \sup_t EI(t)/t = \gamma.$$

In terms of our original variable I_n, we have established that for $a_n = EI_n$ we have the asymptotic relation

$$(1.10) \qquad e^{-\lambda} \sum_{k=0}^{\infty} \lambda^k a_k/k! \sim \gamma \sqrt{\lambda} \quad \text{as } \lambda \to \infty.$$

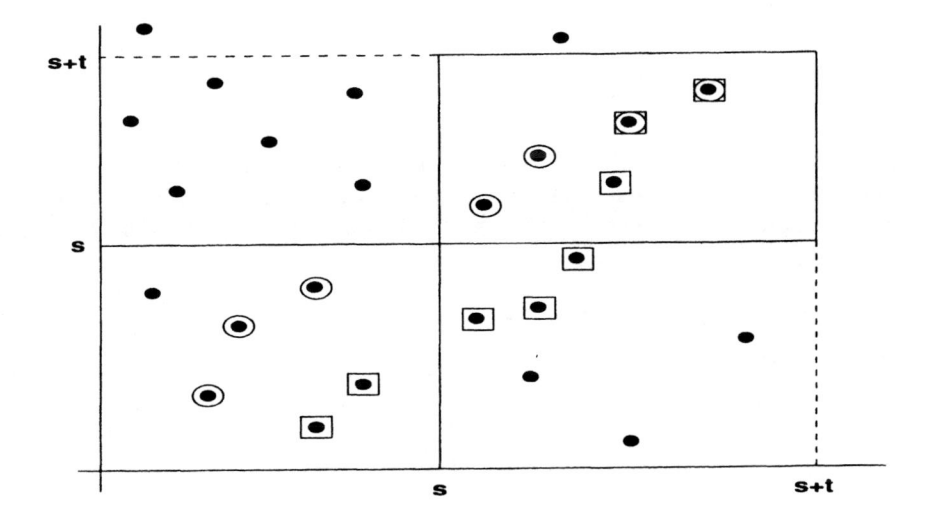

FIG. 1.3. *Increasing sequences illustrating* $I(0,s) + I(s, s+t) \leq I(0, s+t)$.

We would like to conclude that $a_k \sim \gamma k^{1/2}$, and there are general theorems that would permit us to do so without further concern. But since we already know so much about a_k, we can do even better by elementary means.

The fact that the longest increasing subsequence of $\{x_1, x_2, \ldots, x_{n+m}\}$ is not longer than the sum of the lengths of the longest increasing subsequences of $\{x_1, x_2, \ldots, x_n\}$ and $\{x_{n+1}, x_{n+2}, \ldots, x_{n+m}\}$ tells us that for our original sequence $EI_n = a_n$ we have $a_{m+n} \leq a_m + a_n$. This relation is in most ways weaker than (1.9)—by itself, it would only show that $EI_n/n \to 0$—but it does give us useful information.

The expectation bound (1.7) tells us there is a constant C so that $a_m \leq Cm^{1/2}$ for all m; so, in view of the elementary bounds $a_n \leq a_{n-k} + a_k$ and $a_n \geq a_k$ for $n \geq k$, we have a "smoothness" bound for our sequence:

$$(1.11) \qquad |a_n - a_k| \leq C|n-k|^{1/2} \quad \text{for all } n \text{ and } k.$$

Such a smoothness condition gives use a very direct route toward de-Poissonization. If $N(n)$ is a Poisson random variable with mean n, then (1.10) tells us $Ea_{N(n)} \sim \gamma n^{1/2}$ as $n \to \infty$, so we just need to show that $a_n - Ea_{N(n)} = o(n^{1/2})$. An even stronger bound is easily extracted from (1.11), Jensen's inequality, and the fact that Var $N(n) = n$:

$$|a_n - Ea_{N(n)}| \leq \sum_{k=0}^{\infty} |a_n - a_k| \, e^{-n} n^k/k!$$

$$\leq C \sum_{k=0}^{\infty} |n-k|^{1/2} \, e^{-n} n^k/k!$$

$$\leq C \left(\sum_{k=0}^{\infty} (n-k)^2 e^{-n} n^k/k! \right)^{1/4} = Cn^{1/4}.$$

We have thus established that $I_n \sim \gamma n^{1/2}$ as $n \to \infty$, where γ is the constant guaranteed by (1.10).

The constant γ for the long-increasing-subsequence problems has been investigated extensively, and we now know from the work of Logan and Shepp (1977) and Veršik and Kerov (1977) that $\gamma = 2$. The determination of the limiting constant for the increasing-subsequence problem was a genuine tour de force that depended on a rich connection between the increasing-subsequence problem and the theory of Young tableux.

1.5. Flipping Azuma's inequality.

The natural next step in the analysis of I_n is to collect information that tells us about the concentration of I_n about its mean. If we change one of the X_i's in $\{X_1, X_2, \ldots, X_n\}$, then we change $I_n(X_1, X_2, \ldots, X_n)$ by at most one, so if we charge ahead and use the same martingale difference representation (1.5) that proved handy in the long-common-subsequence problem, then we find that we have

$$I_n - EI_n = \sum_{i=1}^n d_i, \quad \text{where} \quad \sum_{i=1}^n \|d_i\|_\infty^2 \leq n.$$

The direct application of Azuma's inequality would just give us

$$P(|I_n - EI_n| \geq t) \leq 2\exp(-t^2/2n),$$

and since $EI_n \sim \gamma\sqrt{n}$, we see that in this case an unthinking application of Azuma's inequality does not get us even as far as a strong law for I_n.

Still, there are some general ideas that often let us push Azuma's inequality a little further, and in the increasing-subsequence problem, these ideas do help us to a considerable extent. The techniques for boosting Azuma's inequality are not free of detail, but they are quite worth learning since they are common to many of the more interesting applications of concentration inequalities.

Before digging into the main argument, we will profit from isolating a widely used trick that is sometimes called "flipping." There is nothing in the lemma to prove, but the statement offers an organizing principle that is often useful.

LEMMA 1.5.1 (flipping lemma). *If we have*

(1.12) $\quad P(Z \geq t) \geq \epsilon \quad \text{and} \quad P(Z \geq u + EZ) < \epsilon,$

then

(1.13) $\quad EZ > t - u \quad \text{and} \quad P(Z \leq t - 2u) \leq P(Z \leq EZ - u).$

The point of the lemma is that if one has a concentration inequality that involves a centering constant like EZ, then a bound on one tail of Z like that of the first inequality of (1.12) can be converted almost automatically into a bound on the other tail of Z like that of the second inequality of (1.13). Moreover, this reversal can be achieved without ever gathering any explicit information about the value of the centering constant. Cognoscienti have been known to express this fact by saying "a concentration inequality lets one do a flip."

THEOREM 1.5.1 (Frieze (1991)). *For any constant $\alpha > \frac{1}{3}$, there is a constant $\beta = \beta(\alpha) > 0$ such that for all sufficiently large n one has*

(1.14) $$P(|I_n - EI_n| \geq n^\alpha) \leq \exp(-n^\beta).$$

Proof. First, for any given $0 < \epsilon < 1$ and $1 \leq n < \infty$, we can define $l = l(\epsilon, n)$ as the unique integer satisfying

(1.15) $$P(I_n < l) < \epsilon \leq P(I_n \leq l).$$

Strange though it may seem, if ϵ is not too terribly small, then $l = l(\epsilon, n)$ can serve as a surrogate for the center of the distribution of I_n.

We then partition $[n] = \{1, 2, \ldots, n\}$ as evenly as possible into $m = \lceil n^b \rceil$ consecutive blocks B_j, $1 \leq j \leq m$, where $0 < b < 1$ will be chosen later. For any $S \subset \{1, 2, \ldots, m\} = [m]$, we then let $I(\cup_{j \in S} B_j)$ denote the cardinality of the longest increasing subsequence of $\{X_i : i \in \cup_{j \in S} B_j\}$. The point of this partitioning is that it lets us introduce a useful variable $Z_n(l)$ that measures how much of $[n]$ can be covered by a union of blocks such that the union does not contain an increasing subsequence that is longer than l:

(1.16) $$Z_n(l) \equiv \max\{|S| : S \subset [m] \text{ and } I(\cup_{j \in S} B_j) \leq l\}.$$

Also, for each integer $0 \leq s \leq m$, we can define a companion variable for $Z_n(l)$ by

(1.17) $$M_n(s) \equiv \max\{I(\cup_{j \in T} B_j) : T \subset [m], \operatorname{card} T = s\}.$$

From the definition of these variables, we have the key inclusion

$$\{I_n \geq l + u\} \cap \{Z_n(l) \geq m - s\} \subset \{M_n(s) \geq u\},$$

which in terms of probabilities gives us for all $u \geq 0$ and $1 \leq s \leq m$,

(1.18) $$P(I_n \geq l + u) \leq P(Z_n(l) \leq m - s) + P(M_n(s) \geq u).$$

We will soon show by "flipping" that Z_n is concentrated near m, but to set up the flip we first need a concentration inequality for Z_n about its mean. To put Z_n into the framework of Azuma's inequality, we define vector random variables V_j by taking the values of X_i for $i \in B_j$ by the order of their index. We can then view $Z_n(l) = Z_n(V_1, V_2, \ldots, V_m)$ as a function of m vector-valued random variables. These variables are independent, though they are not identically distributed—the vectors are not even all necessarily of the same length. Still, the martingale-difference representation of $Z_n(l) - EZ_n(l) = d_1 + d_2 + \cdots + d_m$ works just as before, and because changing one of the V_j's changes Z_n by at most one, we have $\|d_i\|_\infty \leq 1$. Azuma's inequality then tells us that

(1.19) $$P(|Z_n - EZ_n| \geq u) < 2\exp(-u^2/2m).$$

By the definition of $Z_m(l)$, we have the identity

$$P(Z_n(l) = m) \geq P(I_n \leq l) > \epsilon,$$

where the last inequality holds by the definition of $l = l(\epsilon)$.

Now if we introduce a parameter θ and write $u = \sqrt{\theta m}$, then we can choose θ so that $\epsilon = 2\exp(-\theta/2)$, and we have

$$P(Z_n(l) - EZ_n(l) \geq u) < \epsilon \leq P(I_n \leq l) = P(Z_n(l) = m).$$

With these choices, the flipping lemma then tells us that

$$EZ_n(l) \geq m - \sqrt{m\theta}.$$

When we apply this bound on $EZ_n(l)$ in (1.19), we find a bound that measures the closeness of $Z_n(l)$ to m:

$$P(Z_n \leq m - 2\sqrt{m\theta}) \leq \epsilon = 2\exp(-\theta/2).$$

Now we need to look at the second term in (1.18). By Boole's inequality, we have for all $u \geq 0$ that

$$(1.20) \qquad P(M_n(s) \geq u) \leq \binom{m}{s} P(I(\cup_{i=1}^s B_j) \geq u).$$

If we take $s = 2\sqrt{m\theta}$ and $u = 6\sqrt{sn/m}$, so $u > 2e\sqrt{sn/m}$, then the last probability in (1.20) is in the form we need to use the tail bound (1.6). To collect as much information as possible in terms of powers of n, we also let $\theta = n^a$, so with very little arithmetic we find that the right-hand side of (1.20) is bounded by

$$\exp\{s \log m - 2e\sqrt{sn/m}\} \leq \exp\{e(n^{(a+b)/2} \log m - 2n^{\frac{1}{2}+\frac{a}{4}-\frac{b}{4}})\} \equiv \epsilon'.$$

The bottom line is that ϵ' will be exponentially small provided that we have $a + 3b < 2$.

Now what we have to do is put our two estimates together and optimize over our parameter choices to get the best inequality we can. From (1.18), we have

$$P(I_n \geq l + 6\sqrt{sn/m}) \leq \epsilon' + \epsilon,$$

and from the definition of $l(\epsilon)$, we also have

$$P(I_n < l) \leq \epsilon,$$

so if we split the difference and set $l_0 = l + 3\sqrt{sn/m}$, we have a concentration inequality

$$P(|I_n - l_0| \geq 3\sqrt{sn/m}) \leq 2\epsilon + \epsilon'.$$

Now we face the task of choosing a, b, α, and β. We have to meet the naïve constraints $0 < a < 1$ and $0 < b < 1$, and to make ϵ' small, we just need to make sure that

$$(1.21) \qquad a + 3b < 2.$$

Since $3\sqrt{sn/m} \sim n^{\frac{1}{2}+\frac{a}{4}-\frac{b}{4}}$, we want to make $\frac{1}{2} + \frac{a}{4} - \frac{b}{4}$ small. If we choose our parameters so that

$$(1.22) \qquad 0 < \beta < a < \frac{1}{2} + \frac{a}{4} - \frac{b}{4} < \alpha,$$

we see that for all sufficiently large n, we have

(1.23) $$P(|I_n - l_0| \geq n^\alpha) \leq \exp(-n^\beta).$$

If we take b to be just less than $\frac{2}{3}$ and take a to be just above, we see that we have proved the assertion (1.14) of the theorem, except with l_0 in place of EL_n.

To complete the proof, we just need to show that the analogue of (1.23) continues to hold when l_0 is replaced by EL_n. This is easily done by noting that for all k we have

(1.24) $$kP(L_n \geq k) \leq EL_n \leq k + nP(L_n \geq k).$$

In the first inequality of (1.24) we can take $k = l_0 - n^\alpha$ and in the second we can take $k = l_0 + n^\alpha$ to obtain

$$l_0 - n^\alpha - n\exp(-n^\beta) \leq EL_n \leq l_0 + n^\alpha + n\exp(-n^\beta).$$

These bounds and (1.23) suffice to complete the proof of the theorem.

In the crudest terms, the proof of the last theorem might seem to be just the natural result of "applying Azuma's inequality on blocks," but on a deeper look, rather more is afoot. The concentration inequality on Z_n was used to show that a "feasible circumstance" $I_n \leq l$ could be forced to hold with high probability if one restricted the set of observations to which it applies to a set that is a bit smaller than the whole sample. Also, there is an element of pulling oneself up by the bootstraps since we used the crude bound (1.6) to help us handle the Boole terms. Finally, the flipping lemma seems so trivial that one would be right to worry that it is not bringing much to the party, yet the ways the proof can be rewritten all seem to call on some aspect of flipping.

The proof technique of the preceding theorem is worth mastering; but, as developments would have it, better bounds are now available for the tail probabilities of the increasing-subsequence problem—though Frieze's bound did hold the record for a while. We will see in Chapter 6 that Talagrand's isoperimetric inequality gives us an easy way to improve the concentration range from $n^{\frac{1}{3}+\epsilon}$ to $n^{\frac{1}{4}}$.

1.6. Concentration on rates.

Until recently, conventional wisdom held that in a problem where a limit is obtained by subadditivity arguments, one has little hope of saying anything about the rate of convergence. This wisdom continues to hold in the abstract, but concrete problems often provide more handles than just subadditivity. In Alexander (1993), a method was introduced that demonstrates that in many concrete problems of percolation theory, one can supplement subadditivity with a form of superadditivity that can lead to rate results. Alexander (1994b) further applied this philosophy to the long-common-subsequence problem and proved that there is a constant K depending only on the distribution of the letter variables $\{X_i\}$ and $\{Y_i\}$ such that

(1.25) $$n\gamma \geq EL_n \geq \gamma n - K(n \log n)^{1/2}.$$

A beautiful—and highly visual—proof of this fact was given in Rhee (1995). We give Rhee's proof here because it also provides a nice illustration of how concentration inequalities can give information about the value of a mean if one has just a little additional information. The extra information that Rhee uses is given in the next lemma.

LEMMA 1.6.1. *For all integers $n > 0$ and reals $x > 0$, the length L_n of the longest common subsequence of two strings $\{X_1, X_2, \ldots, X_n\}$ and $\{Y_1, Y_2, \ldots, Y_n\}$ of independent and identically distributed letters from a finite alphabet satisfies*

(1.26) $$P(L_{4n} \geq 2x) \leq (4n)^4 P(L_{2n} \geq x)^{1/2}.$$

Proof. We first expand our notation a little bit. If B and B' are integer intervals of $[n]$ (also known as blocks), we write $L_n(B; B')$ for the length of the longest common subsequence of $\{X_i : i \in B\}$ and $\{Y_i : i \in B'\}$. If $L_{4n} \geq 2x$, then we can find two partitions of $[4n]$ into intervals B_i, $1 \leq i \leq 4$, and B'_i, $1 \leq i \leq 4$, such that for all $1 \leq i \leq 4$ we have $L(B_i; B'_i) \geq x/2$. Further, since

$$\text{card } B_1 + \cdots + \text{card } B_4 = 4n$$

and

$$\text{card } B'_1 + \cdots + \text{card } B'_4 = 4n,$$

there is an $1 \leq i \leq 4$ such that $\text{card } B_i + \text{card } B'_i \leq 2n$. Thus we see that if S denotes the set of all pairs of integer intervals B and B' of $[4n]$, where $\text{card } B + \text{card } B' \leq 2n$, then as illustrated in Figure 1.4, we have

(1.27) $$\{L_{4n} \geq 2x\} \subset \bigcup_S \{L_{4n}(B, B') \geq x/2\}.$$

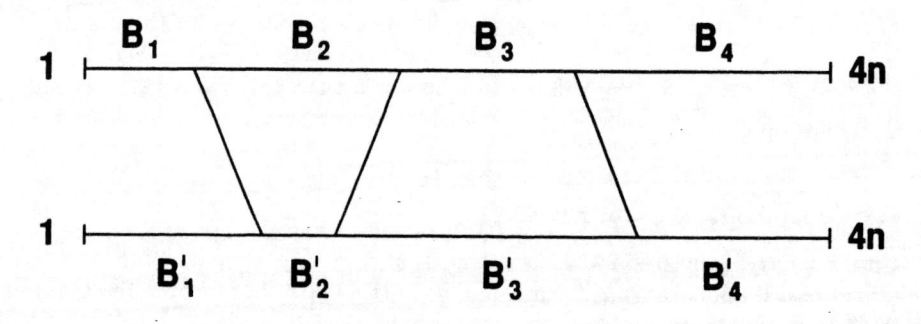

FIG. 1.4. *Double decomposition illustrating 1.27.*

The cardinality of S is easily bounded,

$$\text{card } S \leq \binom{4n}{2}\binom{4n}{2} \leq (4n)^4,$$

so Boole's inequality applied to (1.27) tells us that

(1.28) $$P(L_{4n} \geq 2x) \leq (4n)^4 \max_S P(L_{4n}(B; B') \geq x/2).$$

From the definition of S, we see we can reexpress this maximum,

$$\max_S P(L_{4n}(B; B') \geq x/2) = \max_{s,t: s+t \leq 2n} P(L_{2n}(X_1, \ldots, X_s; Y_1, \ldots, Y_t) \geq x/2),$$

so we just need to bound the last expression in terms of L_{2n}. Now for any integers s and t, we have equality of the distributions

$$L(X_1, \ldots, X_s; Y_1, \ldots, Y_t) \stackrel{d}{=} L(X_1, \ldots, X_t; Y_1, \ldots, Y_s),$$

so for $s + t \leq 2n$ and independent copies $\{X'_i\}$ and $\{Y'_i\}$ of $\{X_i\}$ and $\{Y_i\}$, we have

$$\begin{aligned} P(L_{2n} \geq x) &\geq P(L(X_1, \ldots, X_s, X'_1, \ldots, X'_t; Y_1, \ldots, Y_t, Y'_1, \ldots, Y'_s) > x) \\ &\geq P(L(X_1, \ldots, X_s; Y_1, \ldots, Y_t) > x/2) P(L(X'_1, \ldots, X'_t; Y'_1, \ldots, Y'_s) > x/2) \\ &= P(L(X_1, \ldots, X_s; Y_1, \ldots, Y_t) > x/2)^2, \end{aligned}$$

or, in a line,

$$(1.29) \quad \max_{0 < s+t \leq n} P(L(X_1, \ldots, X_s; Y_1, \ldots, Y_t) > x/2)^2 \leq P(L_{2n} \geq x).$$

On returning the last estimate to (1.28), we see that the lemma is proved.

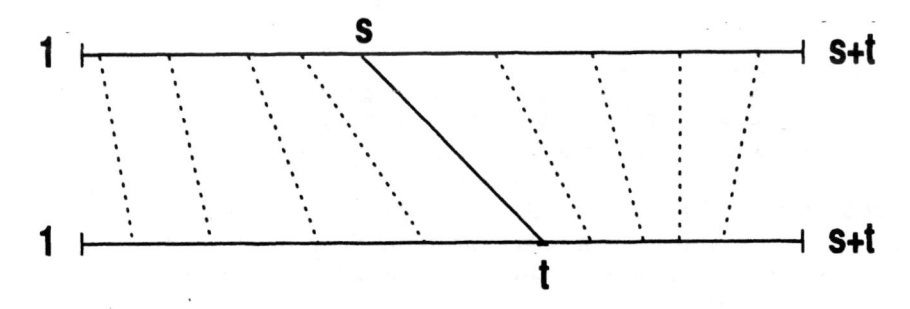

FIG. 1.5. *One of the possibilities contributing to inequality 1.29.*

We now go to the proof of Alexander's inequalities (1.25). The upper bound $\gamma n \geq EL_n$ is an immediate consequence of the superadditivity of the sequence $a_n = EL_n$, so we only need to prove the lower bound. This is where our concentration inequalities make an interesting appearance. When we apply Azuma's inequality to the right-hand side of (1.26), we find for $x \geq EL_{2n}$ that

$$(1.30) \quad P(L_{4n} \geq 2x) \leq 2(4n)^4 \exp(-(x - EL_{2n})^2/16n),$$

and this will turn out to be all that we need to prove Alexander's lower bound.

To put down the details, we first note that for any nonnegative random variable, we have for all $u \geq 0$ that

$$EX \leq u + \int_u^\infty P(X \geq x) \, dx,$$

and by the choice of $X = L_{4n}$ and $u = 2EL_{2n} + K(n \log n)^{1/2}$, we see from (1.30) that the integral contributes negligibly to the upper bound in comparison to u. This proves that we have a K such that

$$\frac{EL_{4n}}{4n} \leq \frac{EL_{2n}}{2n} + K\left(\frac{\log n}{n}\right)^{1/2}.$$

When we apply this inequality with n replaced by $2^k n$, we find

$$\frac{EL_{2^{k+2}n}}{2^{k+2}n} \leq \frac{EL_{2^{k+1}n}}{2^{k+1}n} + \frac{K}{4}\left(\frac{\log 2^k n}{2^k n}\right)^{1/2},$$

and we can sum these inequalities over all $1 \leq k \leq s$ to find

$$\frac{EL_{2^{s+2}n}}{2^{s+2}n} \leq \frac{EL_{2n}}{2n} + K\left(\frac{\log n}{n}\right)^{1/2}.$$

Finally, when we let $s \to \infty$, we have

$$\gamma \leq \frac{EL_{2n}}{2n} + K\left(\frac{\log n}{n}\right)^{1/2},$$

and this is the inequality required by the theorem in the case of even n. The case of odd n follows trivially from the case of even n, so the proof is complete.

1.7. Dynamic programming.

The computation of the length of the longest increasing subsequence or the length of the longest common subsequence can both be handled easily by dynamic programming. In each of these cases, dynamic programming leads almost instantly to an algorithm that runs in time $O(n^2)$ for the problems of size n, but sometimes with a little thought one can do much better.

To see how one version of a dynamic-programming algorithm works for the longest-increasing-subsequence problem, we consider a set of distinct real numbers $\{x_1, x_2, \ldots, x_n\}$, and we construct a labeling of this set as follows:

- First, we label x_1 with the pair $(l_1, l'_1) \equiv (1, 1)$.

- If the set $\{x_1, x_2, \ldots, x_{i-1}\}$ has already been labeled, we form the sets $A = \{l_j : x_j > x_i \text{ with } j < i\}$ and $B = \{l'_j : x_j < x_i \text{ with } j < i\}$.

- If $A \neq \emptyset$ we let $l_i = 1 + \max A$ and otherwise we let $l_i = 1$.

- If $B \neq \emptyset$ we let $l'_i = 1 + \max B$ and otherwise we let $l'_i = 1$.

One can easily check that this provides a labeling where l_i is the length of the longest decreasing subsequence ending with x_i, and l'_i is the length of the longest increasing subsequence ending with x_i. Also, the running time of this algorithm is easily bounded by $O(n^2)$ since we perform n labelings, and each of these has a cost that is clearly bounded by $O(n)$.

One of the amusing aspects of this algorithm is that it provides another proof of the Erdös–Szekeres theorem. To see this, we note that each of the labels is distinct since for any pair of labels (l_i, l_i') and $(1_j, l_j')$ with $i < j$, we have either $l_i < l_j$ or $l_i' < l_j'$. If there is no decreasing sequence larger than a and no increasing subsequence larger than b, then the labels must be contained in the lattice rectangle $[1, a] \times [1, b]$, so since the labels are all distinct, we have $n \leq ab$, completing the proof. There are a good many other proofs of the Erdös–Szekeres theorem, including the original proof. A discussion of these proofs and the directions in which they lead is given in Steele (1995), a survey focused almost exclusively on the Erdös–Szekeres theorem.

To provide an algorithm that more quickly computes the length of the longest increasing subsequence, we follow Fredman (1975). One of the ideas in the speed-up is that we should keep track of only what we need, and we should use a data structure that supports quick implementations of the operations that we need to use.

Fredman builds a table $\{T(j) : j = 1, 2, \ldots\}$ with either a real number or the value NULL stored at the table location $T(j)$ for each j. We begin with $T(1) = x_1$ and the other table values set to NULL. Now for each of the successive values $x_2, x_3, \ldots$, we let k denote the highest address with a non-NULL value and we insert the current value x_j into $T(k+1)$ if $T(k) < x_j$, and otherwise we just put x_j into the $T(m)$ where m is the least integer for which we have $T(m) < x_j$.

From the construction, it should be clear that $L = \max\{k : T(k) \neq \text{NULL}\}$ is equal to the length of the longest common subsequence. For a formal justification, we can just note that for all j we have the truth of the statement "$T(k)$ is the least value of $\{x_1, x_2, \ldots, x_j\}$ for which there is an increasing subsequence of length k terminating with $T(k)$." This *program invariant* is easily verified to remain true after each step of the algorithm.

To see that the algorithm runs in time that is only $O(n \log n)$, we have to add some detail. First, we note that at each stage of the algorithm, the non-NULL values of $T(k)$ are in increasing order, $T(1) < T(2) < \cdots$. Since the table need not be any larger than n and since binary search will let us insert an item into a binary table of size n in time $O(\log n)$, we see that all n items of the set can be placed into the table in time $O(n \log n)$. If we are more careful with our estimates, then we can show that the algorithm will require no more than $n \log n - n \log \log n) + O(n)$ comparisons.

One of the interesting features of Fredman's article is that he also provides a lower bound on the number of comparisons that are needed, and the argument he provides for that lower bound is organized around the probability calculation that in a random sequence the expected number of increasing subsequences of length k is just $\binom{n}{k}/k!$. Using the information-theoretic lower-bound method, Fredman then shows that any algorithm that computes the length of the longest increasing subsequence may require as many as $\frac{1}{2} n \log n + O(n)$ comparisons.

There are not very many problems for which we have algorithms where we know that the order of worst-case running time of the algorithm is equal to the order of a known lower bound for the running time of any algorithm. This is a strong form of efficiency that is almost never found except in algorithms with running times of $O(n)$ where the bound is sometimes trivial or in problems with running times of $O(n \log n))$ where one can use the information-theoretic lower bound.

Algorithms for the long-common-subsequence problem.

Now we should take on the longest-common-subsequence problem, and we will first see how well we do with a straightforward dynamic-programming approach. To give some room, we suppose somewhat more generally than usual that we have strings of letters $x_1, x_2, \ldots, x_m$ and $y_1, y_2, \ldots, y_n$ of possibly unequal length. The dynamic-programming table that we calculate will be an $m \times n$ table where the (j, k) entry $T(i, j)$ of the table will be the length of the longest common subsequence of the prefix strings $x_1, x_2, \ldots, x_i$ and $y_1, y_2, \ldots, y_j$. The dynamic-programming process we consider this time is to fill out the table by rows. Specifically, suppose we have calculated the table entries $T(i, j)$ for all of the first $s - 1$ rows and for the first t entries of row s. We next need to compute $T(s, t + 1)$ just by setting

$$T(s, t+1) = \begin{cases} 1 + T(s-1, t) & \text{if } x_s = y_{t+1}, \\ T(s, t) & \text{if } x_s \neq y_{t+1}. \end{cases}$$

To complete our rule for filling the table, we only need to know how to start a new row, and this is easy. We just set $T(s+1, 1) = 1$ if $x_{s+1} = y_t$ and otherwise we take $T(s+1, 1) = 0$. Since the cost to add a new entry into the table is $O(1)$ and since there are mn table entries, the cost of this easy dynamic-programming algorithm is $O(mn)$ in both time and space.

There is a substantial literature that has been devoted to improving upon this algorithm, and in particular there are careful variations on the dynamic-programming theme given by Hirschberg (1977) that provide an important speedup in the case where the length p of the longest increasing subsequence is either very small or very large in comparison to the length n of the source strings. Hirschberg's algorithms are of running-time orders $O(pn + n \log n)$ and $O((n - p)p \log n)$, respectively, and these can be quite useful in some context; they do not lead to subquadratic algorithms in the cases of greatest concern here.

There are also at least two other approaches to the long-common-subsequence problems that make improvements over the direct application of dynamic programming. In particular, Hunt and Szymansky (1977) provide an $n \log n$ algorithm that is based on a reduction to—of all things—an increasing-subsequence problem. Also, Meyers (1986) has developed an algorithm that is based on Dijkstra's shortest-path algorithm. For more details about these algorithms and their competitors, one should consult the survey of Aho (1990) and the article by Apostolico and Guerra (1987).

1.8. Kingman's subadditive ergodic theorem.

In the preceding sections, we showed using elementary subadditivity and martingale arguments that for the long-common-subsequence-problem and the increasing-subsequence problem, one has the limiting relations

$$L_n/n \to c_{\text{LCS}} \quad \text{and} \quad I_n/\sqrt{n} \to c_{\text{LIS}}.$$

In both cases, the first such relations were proved instead with the assistance of the subadditive ergodic theorem of Kingman (1968). This theorem is one of the most useful results in probability theory, and there are reasons for mentioning it here that go deeper than the need to provide historical perspective. At a minimum, Kingman's theorem tells us immediately that the strong law for the long-common-subsequence problem does not require that the two strings be generated by an independent process; stationarity would certainly suffice.

Kingman's subadditive ergodic theorem has inspired many proofs, even more than the fundamental ergodic theorem of Birkhoff. The proof given here from Steele (1989b) was motivated by the proof of Birkhoff's ergodic theorem given in Shields (1987), where roots of the algorithmic majorization are traced back several additional steps. The idea is that by following a sample path and collecting information from the times when the average is near its limit inferior, one can use subadditivity to bound the limsup. Given the richness and importance of Kingman's theorem, there is a certain charm to building its proof on a principle that goes all the way back to Fekete's lemma. Kingman's original proof and most others that have followed were built on much different principles.

THEOREM 1.8.1. *If T is a measure-preserving transformation of the probability space $(\Omega, \mathcal{F}, \mu)$ and $\{g_n, 1 \leq n < \infty\}$ is a sequence of integrable functions satisfying*

(1.31) $$g_{n+m}(x) \leq g_n(x) + g_m(T^n x),$$

then, with probability one, we have

$$\lim_{n \to \infty} g_n(x)/n = g(x) \geq -\infty,$$

where $g(x)$ is an invariant function for T.

Proof. If we define a new process g'_m by

$$g'_m(x) = g_m(x) - \sum_{i=1}^{m-1} g_1(T^i x),$$

then $g'_m(x) \leq 0$ for all x, and g'_m again satisfies (1.31). Since Birkhoff's ergodic theorem can be applied to the second term of g'_m, the almost sure convergence of g'_m/m implies the almost sure convergence of g_m/m. Thus, without loss of generality, we can assume $g_m(x) \leq 0$.

Next, we check that $g(x) = \liminf_{n \to \infty} g_n(x)/n$ is an invariant function. By (1.31), we have $g_{n+1}(x)/n \leq g_1(x)/n + g_n(Tx)/n$, so taking the limit inferior, we see that $g(x) \leq g(Tx)$. Thus we find $\{g(x) > \alpha\} \subset T^{-1}\{x : g(x) > \alpha\}$, and since T is measure preserving, the sets $\{g(x) > \alpha\}$ and $T^{-1}\{g(x) > \alpha\}$ differ by at most a set of measure zero. Consequently, $g(Tx) = g(x)$ almost surely, and g is measurable with respect to the invariant σ-field $\mathcal{A}$. Thus we can also assume without loss of generality that for all $x \in \Omega$ we have $g(T^k x) = g(x)$ for all integers k.

Now, given $\epsilon > 0$, $1 < N < \infty$, and $0 < M < \infty$, we let $G_M(x) = \max\{-M, g(x)\}$ and consider the set

$$B(N, M) = \{x : g_l(x) > l(G_M(x) + \epsilon) \text{ for all } 1 \leq l \leq N\}$$

and its complement $A(N, M) = B(N, M)^c$. For any $x \in \Omega$ and $n \geq N$, we then decompose the integer set $[1, n)$ into a union of three classes of intervals by the following algorithm:

Begin with $k = 1$. If k is the least integer in $[1, n)$ which is not in an interval already constructed, then consider $T^k x$. If $T^k x \in A(N, M)$, then there is an $l \leq N$ so that $g_l(T^k x) \leq l(G_M(T^k x) + \epsilon) = l(G_M(x) + \epsilon)$, and if $k + l \leq n$, we take $[k, k + l)$ as an element of our decomposition. If $k + l > n$, we just take the singleton interval $[k, k+1)$, and finally, if $T^k x \in B(N, M)$, we also take the singleton $[k, k+1)$.

Thus for any x, we have a decomposition of $[1, n)$ into a set of u intervals $[\tau_i, \tau_i + l_i)$, where $g_{l_i}(T^{\tau_i} x) \leq l_i(G_M(x) + \epsilon)$ with $1 \leq l_i \leq N$, and two sets of singletons: one set of v singletons $[\sigma_i, \sigma_i + 1)$ for which $1_{B(N,M)}(T^{\sigma_i} x) = 1$ and a second set of w singletons contained in $(n - N, n)$.

By the fundamental subadditive inequality (1.31), our decomposition of $[1, n)$, the invariance of g, and the assumption that $g_m(x) \leq 0$, we have the bounds

$$g_n(x) \leq \sum_{i=1}^{u} g_{l_i}(T^{\tau_i} x) + \sum_{i=1}^{v} g_1(T^{\sigma_i} x) + \sum_{i=1}^{w} g_1(T^{n-i} x)$$

(1.32)
$$\leq (G_M(x) + \epsilon) \sum_{i=1}^{u} l_i \leq G_M(x) \sum_{i=1}^{n} l_i + n\epsilon.$$

Also, by the construction of the covering intervals, we have

(1.33)
$$n - \sum_{k=1}^{n} 1_{B(N,M)}(T^k x) - N \leq \sum_{i=1}^{u} l_i,$$

so by (1.33) and Birkhoff's theorem, we find

(1.34)
$$\liminf_{n \to \infty} n^{-1} \sum_{i=1}^{n} l_i \geq 1 - E(1_{B(N,M)} \mid \mathcal{A}) \quad \text{a.s.}$$

By (1.32), we then conclude for any $K \leq M$ that

$$\limsup_{n \to \infty} g_n(x)/n \leq G_M(x)(1 - E(1_{B(N,M)} \mid \mathcal{A})) + \epsilon \quad \text{a.s.}$$

(1.35)
$$\leq G_K(x)(1 - E(1_{B(N,M)} \mid \mathcal{A})) + \epsilon \quad \text{a.s.}$$

Now, the definition of $B(N, M)$ guarantees that $1_{B(N,M)} \to 0$ almost surely as $N \to \infty$ for any fixed K, so letting $N \to \infty$ in (1.35) implies

$$\limsup_{n \to \infty} g_n(x)/n \leq G_K(x) + \epsilon \quad \text{a.s.}$$

Since the last inequality holds for all $K \geq 0$ and $\epsilon > 0$, we see that

$$\limsup_{n \to \infty} g_n(x)/n \leq \liminf_{n \to \infty} g_n(x)/n \quad \text{a.s.}$$

and the proof of Kingman's theorem is complete.

1.9. Observations on subadditive subsequences.

Subadditive sequences are such handy tools that one is almost forced to explore the extent to which the subadditivity hypotheses can be modified without harm. Two particularly natural modifications are (1) to relax the requirement that subadditivity holds for all m and n and (2) to relax the subadditivity condition itself to a weaker inequality. Marvelously, these two possibilities are closely related, and in both cases, one can provide results that are the best possible.

THEOREM 1.9.1 (DeBruijn–Erdös (1952a)). *If the sequence $\{a_n\}$ of real numbers satisfies the subadditivity condition*

$$a_{m+n} \leq a_m + a_n \quad \text{over the restricted range} \quad \frac{1}{2}n \leq m \leq 2n,$$

then $\lim_{n \to \infty} a_n/n = \gamma$, where $-\infty \leq \gamma < \infty$ and $\gamma = \inf a_n/n$.

Proof. If we set $g(n) = a_n/n$, then subadditivity expresses itself as a convexity relation:

$$(1.36) \quad g(n) \leq kn^{-1}g(k) + (n-k)n^{-1}g(n-k) \quad \text{for all} \quad \frac{1}{2} \leq k^{-1}(n-k) \leq 2.$$

The idea that drives the proof is that when n is chosen so that $g(n)$ is large, then (1.36) can be used to show $g(k)$ is large for many values of k.

To exploit this idea, we choose a subsequence n_τ so that as $\tau \to \infty$, we have $g(n_\tau) \to g^* = \limsup g(n)$. We then have for any $\epsilon > 0$ that there is an $N(\epsilon)$ such that $j \geq N(\epsilon)$ implies $g(j) \leq g^* + \epsilon$, and we can even require of $N(\epsilon)$, that for $n_\tau > N(\epsilon)$, we also have $g(n_\tau) \geq g^* - \epsilon$.

For k such that $n_\tau \leq 3k \leq 2n_\tau$, where $n_\tau \geq 3N(\epsilon)$, we have $n_\tau - k \geq N(\epsilon)$, so inequality (1.36) implies

$$g^* - \epsilon \leq g(n_\tau) \leq kg(k)/n_\tau + (n_\tau - k)(g^* + \epsilon)/n_\tau.$$

This bound simplifies to $g^* - 2n_\tau\epsilon/k \leq g(k)$, and hence for $n_\tau \geq 3N(\epsilon)$, we have

$$g^* - 6\epsilon \leq \min\{g(k) : n_\tau \leq 3k \leq 2n_\tau\}.$$

Next, we let $g_* = \liminf_{n \to \infty} g(n)$ and choose m so that $g(m) \leq g_* + \epsilon$. Taking $a_1 = m$, $a_2 = 2m$, and $a_{k+1} = a_k + a_{k-1}$ for $k \geq 3$, we have by (1.36) and induction that $g(a_k) \leq g_* + \epsilon$. We also have $a_{k+1}/a_k \leq \frac{3}{2}$ for $k \geq 3$ since $a_{k+1}/a_k \leq 1 + a_{k-1}/a_{k+1}$ and $a_{k+1} \geq 2a_{k-1}$.

If there is no element of $A = \{a_1, a_2, \ldots\}$ that is also in $\{j : n_\tau \leq 3j \leq 2n_\tau\}$, then there would be a k such that

$$a_{k+1}/a_k > \frac{(2n_\tau)/3}{(3n_\tau)/3} = \frac{2}{3};$$

hence we therefore find $\{j : n_\tau/3 \leq j \leq 2n_\tau/3\} \cap A \neq \phi$. By (1.36) and the fact that for all k we have $g(a_k) \leq g_* + \epsilon$, we find $g^* - 6\epsilon \leq g_* + \epsilon$. Since this inequality holds for all $\epsilon > 0$, we conclude that $g^* = g_*$.

A couple of easy points serve to illuminate this theorem. First, it is not hard to check that the hypothesis cannot be relaxed to $n/\alpha \leq m \leq \alpha n$ for any $\alpha < \frac{1}{2}$. Further, one should note that there is a three-term version of Theorem 1.9.1 where we require $f(n_1 + n_2 + n_3) \leq f(n_1) + f(n_2) + f(n_3)$ for all n_i satisfying $\frac{1}{3} \leq n_i/n_j \leq 3$. Here one should note that for the restricted subadditivity theorem, the three-term result is not contained in the two-term result, but in the basic unrestricted case, the three-term version is just a special case of the two-term version.

The next theorem shows that the basic subadditivity condition can be relaxed in terms of quality as well as extent without doing damage to the conclusion. The theorem also illustrates the gains that are made by relaxing the range over which one needs the subadditive inequality. In many circumstances, it can be difficult (or impossible) to show subadditivity of a sequence over its whole range, and the proof of the next result illustrates that at the very least it can be much more convenient to build sequences that have subadditivity over a restricted set.

THEOREM 1.9.2 (DeBruijn–Erdős (1952a)). *Suppose ϕ is a positive and nondecreasing function that satisfies*

$$\int_1^\infty \frac{\phi(t)}{t^2} dt < \infty.$$

If $\{a_n\}$ satisfies the relaxed subadditivity relation

$$a_{n+m} \leq a_n + a_m + \phi(n+m) \quad \text{for } \frac{1}{2}n \leq m \leq 2n,$$

then as $n \to \infty$, a_n/n converges to $\gamma = \inf a_n/n$.

Proof. The natural idea is to add a term b_n to a_n so that the sum $a_n + b_n$ is subadditive. For this idea to lead to the convergence of a_n/n, one also needs for b_n to be small in the sense that $b_n = o(n)$. We will verify that a suitable choice is

$$b_n = 3n \int_n^\infty \phi(3t) t^{-2} dt.$$

By the convergence of the integral of $\phi(x)/t^2$, the added term is $o(n)$, so we just need to check the subadditivity of $c_n = a_n + b_n$. From the monotonicity of ϕ, we have the estimation

$$\int_a^b \phi(3t) t^{-2} dt \geq \phi(3a) \int_a^b t^{-2} dt = \phi(3a)\{1/a - 1/b\},$$

so substituting into the definition of $\{c_n\}$, we see that

$$c_{m+n} - c_m - c_n$$
$$= a_{n+m} - a_n - a_m - 3m \int_m^{n+m} \phi(3t) t^{-2} dt - 3n \int_n^{n+m} \phi(3t) t^{-2} dt$$
$$\leq \phi(n+m) - 3m\phi(3m)\{1/m - 1/(n+m)\} - 3n\phi(3n)\{1/n - 1/(n+m)\}.$$

By the range restriction $\frac{1}{2}n \leq m \leq 2n$ and the monotonicity of ϕ, we have

$$\phi(n+m) \leq \phi(3m) \quad \text{and} \quad \phi(n+m) \leq \phi(3n),$$

so we have that $c_{m+n} - c_m - c_n$ is bounded above by

$$\phi(3m)\{1 - 3m(1/m - 1/(n+m))\} + \phi(3n)\{1 - 3n(1/n - 1/(n+m))\} \leq 0.$$

There are many embellishments of the subadditive limit relation that might deserve to be developed at this point, but one stands out above the rest for the guidance that it offers in problems where one needs information on the rate of convergence of a_n/n to its limit. The central idea is that if one has *both* a subadditive relation and a superadditive relation, then a rate result of some type is guaranteed. One of the oldest and nicest of such results is due to Pólya and Szegö (1924).

LEMMA 1.9.1 (Subadditive Rate Result). *If a real sequence $a_1, a_2, a_3, \ldots$ satisfies*

(1.37) $\quad a_m + a_n - 1 < a_{m+n} < a_m + a_n + 1 \quad \text{for all } m, n = 1, 2, 3, \ldots,$

then there is a finite constant ω such that

$$|a_n/n - \omega| < 1 \quad \text{for all } n.$$

Proof. The proof uses a different principle than the convexity and limit-supremum ideas of the previous results, and this difference is suggestive of how one might proceed in other problems where a two-sided condition is available.

We first note that from $2a_m - 1 < a_{2m} < 2a_m + 1$, we get a bound on the change from doubling the index:

(1.38) $$\left| \frac{a_{2m}}{2m} - \frac{a_m}{m} \right| < \frac{1}{2m}.$$

We then note that we have convergence of the series

$$\frac{a_1}{1} + \left(\frac{a_2}{2} - \frac{a_1}{a_1}\right) + \left(\frac{a_4}{4} - \frac{a_2}{a_2}\right) + \left(\frac{a_8}{8} - \frac{a_4}{a_4}\right) + \cdots = \lim_{n \to \infty} \frac{a_{2^n}}{2^n} = \omega$$

since in view of (1.38), it is majorized by

$$|a_1| + 2^{-1} + 2^{-2} + 2^{-3} + \cdots.$$

By the immediately preceding theorem, we already know that $a_n/n \to \inf a_n/n$, so we can identify the subsequence limit ω as $\inf a_n/n$. Finally, we note that by telescoping and (1.38), we have

$$\left| \omega - \frac{a_m}{m} \right| < \left| \frac{a_{2m}}{2m} - \frac{a_m}{m} \right| + \left| \frac{a_{4m}}{4m} - \frac{a_{2m}}{2m} \right| + \cdots < \frac{1}{2m} + \frac{1}{4m} + \frac{1}{8m} \cdots = \frac{1}{m}.$$

Naturally, the original proof of this lemma did not call on the theorem of DeBruijn and Erdös since that result came almost thirty years later. Pólya and Szegö (1924) instead used a very clever interpolation argument to extract convergence of a_n/n from the fact that one has convergence along the subsequence $n = 1, 2, 4, 8, 16, \ldots$.

1.10. Additional notes.

1. Useful surveys of the theory and practice of sequence comparisons are given in the book of Sankoff and Kruskal (1983) and the article of Waterman (1984).

2. The techniques built on the exploitation of inequalities like that of Azuma (1967) are often referred to as "bounded-difference methods." In an early combinatorial example, Maurey (1979) exploited martingales with bounded differences to prove an isoperimetric inequality for the set of permutations. The bounded-difference method is naturally part of the more general theme of concentration inequalities, and by the time of the volume by Milman and Schechtman (1986), such inequalities were understood to be at the heart of many questions in the geometry and the probability theory of Banach spaces. The central idea of Azuma's inequality goes back at least to Hoeffding (1963), and the later article still seems surprisingly fresh and original.

The method of bounded differences became an essential part of the tool kit for combinatorists and computer scientists only after the compelling applications made by Shamir and Spencer (1987) to chromatic numbers in random graphs and by Rhee and Talagrand (1987) to the traveling-salesman problem. Bollobás (1987, 1990) and McDiarmid (1989) give useful surveys that illustrate the a range of the bounded-difference method to problems in combinatorics and combinatorial optimization. There are also many nice examples in the beautiful book by Alon, Spencer, and Erdös (1992). The number and depth of new applications continues to build, and the "concentration of measure" phenomenon is emerging as one of the central ideas of probability theory—or perhaps of analysis.

More recently, Talagrand (1995) introduced a class of isoperimetric inequalities for product spaces that has something important to add in almost all of the problems where the bounded-difference method has made a contribution. It may not be fair to say that Talagrand's method makes techniques based on Azuma's inequality obsolete, but the statement is close enough to the truth to make any application of Azuma's inequality a candidate for quick improvement.

3. There is an extensive literature on the increasing-subsequence problem that is surveyed in Steele (1995). One engaging variation on the problem asks how well one can do if the choices for the subsequence are made sequentially without the benefit of knowing the values of the observations that follow in the sequence. If u_n denotes the expected value of the best one can do with any selection strategy that one applies sequentially to independent uniform variables $\{X_1, X_2, \ldots, X_n\}$, then the problem is to determine the asymptotic behavior of u_n. Samuels and Steele (1981) solved the problem by showing that $u_n \sim \sqrt{2n}$ as $n \to \infty$. One interpretation of this result is that u_n represents how well a nonclairvoyant individual can perform, while the problem studied earlier in this chapter tells how well a "prophet" (or clairvoyant individual) could expect to perform. The net result is that the prophet can be expected to do better in the long run *only by a factor of* $\sqrt{2}$.

4. The long-common-subsequence problem has a natural reformulation as a first-passage problem in percolation theory, and Alexander (1995c) develops rate

results for such problems that can be used to provide a complementary approach to the rate results developed in this chapter.

5. Pólya and Szegö (1924) give a crystal-clear statement of the basic lemma on subadditive sequences in Problem 98, Section 1, and they cite Fekete (1923) as their source. Perhaps Pólya and Szegö have shown more than the usual generosity in their citation of their friend and fellow Hungarian. Fekete does give the basic argument for the lemma in his article, but there is not any independent statement of the lemma in his work. There is also no development that suggests that Fekete viewed his argument as more than expediency of the moment. In contrast, Pólya and Szegö seem to have valued this little gem, and they tease out several closely related results.

Still, history is never clear, and there is no way to be sure that Pólya and Szegö provided the critical path by which the lemma on subadditive sequences has become part of the mathematical consciousness. Hille and Phillips (1957) devote an eighteen-page chapter to subadditive functions, the continuous-variable analogue of our subadditive sequences. They cite uses of subadditivity that go back to Minkowski's characterization of convex sets and Ch.-J. de la Vallée-Poussin's theory of moduli of continuity. They also examine subadditive functions under minimal assumptions on regularity and provide information about functions that are only subadditive on a bounded open interval. But Hille and Phillips never mention either Fekete or Pólya and Szegö.

Finally, Hammersley (1962) provides a generalization of Fekete's lemma that is closely related to the theorems of DeBruijn and Erdös (1952a) but which calls on a much different method of proof. Hammersley's work was motivated by percolation theory, where subadditivity is a bread-and-butter technique.

6. The Poissonization device has a long history, and so does the necessity of de-Poissonization. The key benefit of the device is that it brings exact independence into problems where one would otherwise have to fight with some sort of approximate independence. The price one has to pay is that a de-Poissonization argument is needed to get back to the original problem, but the detour still almost always provides a great saving of technical analysis.

Aldous (1989, p. 107) gives an engaging heuristic principle that goes far to suggest when de-Poissonization is possible, but Aldous's principle is focused on small probabilities rather than large expected values. for the problems considered here, the right principle seems to be simply that de-Poissonization is easy whenever we have a rough growth condition on a_n and some measure of smoothness. Any de-Poissonization argument is in fact a Tauberian theorem for the Borel mean, and there is a substantial battery of such theorems ready for use off the shelf. Bingham (1981) gives a modern development of Tauberian theorems that is pleasantly accessible to probabilists. An example of how one can use a differentiation technique to make a series more amenable to the traditional hypotheses of Tauberian theory is given in Hochbaum and Steele (1982).

CHAPTER 2

Concentration of Measure and the Classical Theorems

This chapter begins our study of the Euclidean traveling-salesman problem (TSP) and a closely associated collection of basic objects of geometric combinatorial optimization, including the minimal spanning tree and minimal matchings. The central topic of the chapter is the Beardwood–Halton–Hammersley (BHH) theorem, which provides the most basic information on the large-sample behavior of the TSP. The chapter makes more sophisticated uses of subadditivity than were found in the first chapter, and it adds to our martingale toolkit. The chapter also introduces the space-filling curve heuristic for the TSP and gives the first indications of the general analytical uses for space-filling curves.

2.1. The TSP and a quick application of Azuma's inequality.

Perhaps the most studied problem in the theory of combinatorial optimization is the traveling-salesman problem. In the most explicit case, this problem calls for the calculation of the shortest tour through n points $\{x_1, x_2, \ldots, x_n\} \subset \mathbb{R}^d$, that is, the calculation of a permutation σ that minimizes

$$|x_{\sigma(1)} - x_{\sigma(2)}| + |x_{\sigma(2)} - x_{\sigma(3)}| + \cdots + |x_{\sigma(n-1)} - x_{\sigma(n)}| + |x_{\sigma(n)} - x_{\sigma(1)}|,$$

where $|x - y|$ denotes the usual Euclidean distance from x to y.

From a probabilistic point of view, the natural place to begin the study of the TSP is with the analysis of the length of the shortest tour through a random sample $\{X_1, X_2, \ldots, X_n\}$ of the unit d-cube $[0,1]^d$. If we denote this length by $L_n(X_1, X_2, \ldots, X_n)$, we will find that there is a benefit to beginning the study of L_n by looking at the implications of the martingale representation and of Azuma's inequality. The historical development went by another path, but from the current perspective, one of the most striking features of L_n is the extent to which it is concentrated about its mean.

As usual, we represent the centered variable $L_n - EL_n$ as the sum of martingale differences d_i that are in turn written as

$$d_i = E(L_n(X_1, X_2, \ldots, X_i, \ldots, X_n) - L_n(X_1, X_2, \ldots, \hat{X}_i, \ldots, X_n) \,|\, \mathcal{F}_i),$$

where $\{\hat{X}_i\}$ is an independent sequence of random variables with the same distribution as the $\{X_i\}$'s. This time, one needs to be a little trickier to get good

L_∞ bounds on the d_i's. From the fact that no two points of the d-cube can be separated by no more than a distance of $\sqrt{d}$, one can easily check that for all i we have $\|d_i\|_\infty \leq 2\sqrt{d}$. We will need a more serious bound to make real progress with Azuma's inequality.

To move toward a useful bound on $\|d_i\|_\infty$, we first focus on the absolute difference

$$\Delta_{n,i} \equiv |L_n(X_1, X_2, \ldots, X_i, \ldots, X_n) - L_n(X_1, X_2, \ldots, \hat{X}_i, \ldots, X_n)|.$$

For any finite set S and any point x, the TSP functional satisfies

$$L(S) \leq L(S \cup \{x\}) \leq L(S) + 2 \min_{y \in S} |x - y|,$$

and if we apply this bound to the set $S = \{X_1, X_2, \ldots, X_i, \ldots, X_n\} \setminus \{X_i\}$ with both $x = X_i$ and $x = \hat{X}_i$, we find a very natural bound on $\Delta_{n,i}$,

(2.1) $$\Delta_{n,i} \leq 2 \min_{j: j \neq i} |\hat{X}_i - X_j| + 2 \min_{j: j \neq i} |X_i - X_j|,$$

but for the purpose of bounding the $\|d_i\|_\infty$'s we will actually use a cruder bound,

(2.2) $$\Delta_{n,i} \leq 2 \min_{j: j > i} |\hat{X}_i - X_j| + 2 \min_{j: j > i} |X_i - X_j|.$$

Now if we write

(2.3) $$g_m(x) \equiv E\left(\min_{1 \leq i \leq m} |X_i - x|\right),$$

then after taking due note of dummy indices, we have

$$|d_i| \leq 2E\left(\min_{j: j > i} |X_i - X_j| + \min_{j: j > i} |\hat{X}_i - X_j| \,\bigg|\, \mathcal{F}_i\right)$$

(2.4) $$= 2g_{n-i}(X_i) + 2E g_{n-i}(\hat{X}_i).$$

To bound the last two terms, we need a geometric lemma, and to accommodate some later needs, we will prove a result that is a bit more general than is needed right now.

LEMMA 2.1.1. *If X_i, $1 \leq i < \infty$, are independent and uniformly distributed in $[0, 1]^d$ and $x \in [0, 1]^d$, then for any $0 < p < \infty$, we have*

(2.5) $$g_{m,p}(x) \equiv E\left(\min_{1 \leq i \leq m} |X_i - x|^p\right) \leq \frac{p}{d} \Gamma\left(\frac{p}{d}\right) 2^p \omega_d^{-p/d} m^{-p/d},$$

where ω_d is the volume of the ball of unit radius in $\mathbb{R}^d$.

Proof. If $B(x, \lambda)$ is the ball of radius λ about x, where $0 < \lambda < 1$, then the volume of $B(x, \lambda) \cap [0, 1]^d$ is minimized when x coincides with one of the corners of $[0, 1]^d$, and in that case $B(x, \lambda) \cap [0, 1]^d$ has volume equal to $\omega_d \lambda^d 2^{-d}$, where ω_d is the volume of the unit ball in $\mathbb{R}^d$. We therefore find that for any $x \in [0, 1]^d$, we have

(2.6) $$P\left(\min_{1 \leq i \leq m} |X_i - x| \geq \lambda\right) \leq (1 - \omega_d \lambda^d 2^{-d})^m \leq \exp(-m \omega_d \lambda^d 2^{-d}).$$

Finally, since

(2.7) $$E\left(\min_{1\leq i\leq m}|X_i-x|^p\right)=p\int_0^\infty \lambda^{p-1}P\left(\min_{1\leq i\leq m}|X_i-x|\geq \lambda\right)d\lambda,$$

we can apply (2.6) to the integrand of (2.7) and complete the proof by recalling the general integral

$$\int_0^\infty \lambda^a e^{-b\lambda^c}d\lambda = c^{-1}b^{-(a+1)/c}\Gamma((a+1)/c).$$

When we apply (2.5) to the simpler $p=1$ version of $g_{m,p}$ that is given in (2.3) and used in (2.4), we find the desired bound on $\|d_i\|_\infty$,

(2.8) $$\|d_i\|_\infty \leq c(d)(n-i+1)^{-1/d} \quad \text{for } d \geq 2;$$

and, in the same way, we can calculate from (2.1) and (2.5) for general $p \geq 1$ and $m=n-1$ that

(2.9) $$\|d_i\|_p \leq c(d)(p/n)^{1/d} \quad \text{for } d \geq 2.$$

The constants $c(d)$ that are used in (2.8) and (2.9) depend only on the dimension and so we have no need to be explicit with regard to them. In fact, from this point on, we will often use c as a generic constant that may change from one expression to the next. When we expect a constant to have a less ephemeral life, we will use either Greek letters or capitals.

The critical calculation for the application of Azuma's lemma is that from (2.8) and the usual estimates of sums, we have

(2.10) $$\sum_{i=1}^n \|d_i\|_\infty^2 \leq \begin{cases} c\log n & \text{for } d=2, \\ cn^{(d-2)/d} & \text{for } d>2, \end{cases}$$

where for the first time we take advantage of our convention about the generic constants c. By Azuma's inequality, we therefore find that we have

$$P(|L_n - EL_n| \geq t) \leq \begin{cases} \exp(-ct^2/\log n) & \text{for } d=2, \\ \exp(-ct^2/n^{(d-2)/d}) & \text{for } d>2. \end{cases}$$

This inequality tells us that L_n is concentrated about its mean, and shortly we will see that one can extract from these tail estimates more than is needed to get a strong law for the TSP. For the moment, we just record that even the most simple-minded application of the Borel–Cantelli lemma would give us

(2.11) $$L_n - EL_n = O(\log n) \quad \text{a.s.} \quad \text{for } d=2$$

and

(2.12) $$L_n - EL_n = O(n^{(d-2)/2d}\sqrt{\log n}) \quad \text{a.s.} \quad \text{for } d>2.$$

One can sharpen the last two estimates a bit by traditional methods, but now is not the time for such arguments since the tail bounds that we have obtained by direct use of Azuma's inequality are not the best that are now known. Much of the more recent work on the TSP has focused on techniques for improving on the Azuma-based bounds, and we will study those techniques extensively in subsequent chapters. For the moment, a more pressing—and more elementary—task is to develop some information about the size of the tour length. In particular, we need to determine the asymptotic behavior of the expected tour length.

2.2. Easy size bounds.

First, we should get a basic idea of the size of EL_n for the TSP in $[0,1]^d$. If $\{x_1, x_2, \ldots, x_n\}$ is any finite subset of $[0,1]^d$, the fact that $[0,1]^d$ can be covered with $O(n)$ boxes with diameter $O(n^{-\frac{1}{d}})$ tells by an easy pigeonhole argument that there is a constant $c = c(d)$ such that

$$(2.13) \qquad \min\{|x_i - x_j| : x_i, x_j \in \{x_1, x_2, \ldots, x_n\}\} \leq cn^{-\frac{1}{d}}.$$

If l_n denotes the length of the longest TSP tour of any n-point subset of $[0,1]^d$, then the last bound tells us that

$$l_n \leq l_{n-1} + 2cn^{-\frac{1}{d}}.$$

By summing this recursion, we see that (for another c) we have for the TSP in $[0,1]^d$ that

$$(2.14) \qquad \|L_n\|_\infty \leq cn^{(d-1)/d}.$$

As luck would have it, the right order of the mean EL_n is actually $O(n^{(d-1)/d})$ since by a computation that perfectly parallels the proof of Lemma 2.1.1, we have for $\{X_i\}$ independent and uniformly distributed in $[0,1]^d$ that there is a $c > 0$ such that

$$(2.15) \qquad E\min\{|X_i - X_j| : X_i, X_j \in \{X_1, X_2, \ldots, X_n\}\} \geq cn^{-\frac{1}{d}}.$$

Because any tour through $\{X_1, X_2, \ldots, X_n\}$ contains exactly n edges and since any edge must be at least as large as $\min\{|X_i - X_j| : X_i, X_j \in \{X_1, X_2, \ldots, X_n\}\}$, inequality (2.15) entails

$$(2.16) \qquad EL_n \geq cn^{(d-1)/d},$$

even with the same $c > 0$ as in (2.15).

Since we have both upper and lower bounds on EL_n that are of order $n^{(d-1)/d}$, the natural goal is an exact asymptotic. This goal turns out to be easy enough since we have already developed most of the techniques that we will need.

2.3. Another mean Poissonization.

For any set $S = \{x_1, x_2, \ldots, x_n\} \subset \mathbb{R}^d$, we will write $L(S)$ to denote the length of the shortest tour through S. We then let Π denote the Poisson process in $\mathbb{R}^d$ with unit intensity and define a new stochastic process $\{Z(t) : t \geq 0\}$ by

$$Z(t) = L(\Pi[0,t]^d).$$

We will eventually want to extract the asymptotics of $a_k = EL_k$ from $EZ(t)$, so we should first take account of some basic facts. The cardinality of the set $\Pi[0,1]^d$ is a Poisson random variable with mean t^d, so by scaling, $EZ(t)$ can be expressed in terms of a_k by

$$(2.17) \qquad EZ(t) = t\sum_{k=0}^{\infty} a_k e^{-t^d} t^{dk}/k!.$$

We should also record here that since $a_k = O(n^{(d-1)/d})$ by our L^∞ bound on L_n, we see that $EZ(t)$ is a continuous (even a real analytic) function of t. We will proceed to show that $EZ(t)/t^d$ converges to a constant, and we will use a method that is just a modest extension of the subadditive arguments that were useful in the first chapter.

The key observation is that if $\{Q_i\}$, $1 \leq i \leq m^d$, is a partition of $[0,t]^d$ into cubes of edge length t/m, then the tours of the sets $S_i = \{x_1, x_2, \ldots, x_n\} \cap Q_i$ can be joined together as in Figure 2.1 to form a tour through $\{x_1, x_2, \ldots, x_n\} \cap [0,t]^d$. This knitting process incurs an incremental cost of only $O(m^{d-1})$ since we choose one point out of each of the m^d subtours and since that there is a tour through these m^d chosen points that has length bounded by $O(m^{d-1})$ by our L^∞ bound. The bottom line is that there exists a constant C such that

$$(2.18) \quad L(\{x_1, x_2, \ldots, x_n\} \cap [0,t]^d) \leq \sum_{i=1}^{m^d} L(\{x_1, x_2, \ldots, x_n\} \cap Q_i) + Ctm^{d-1}$$

for all integers $m \geq 1$ and real $t \geq 0$.

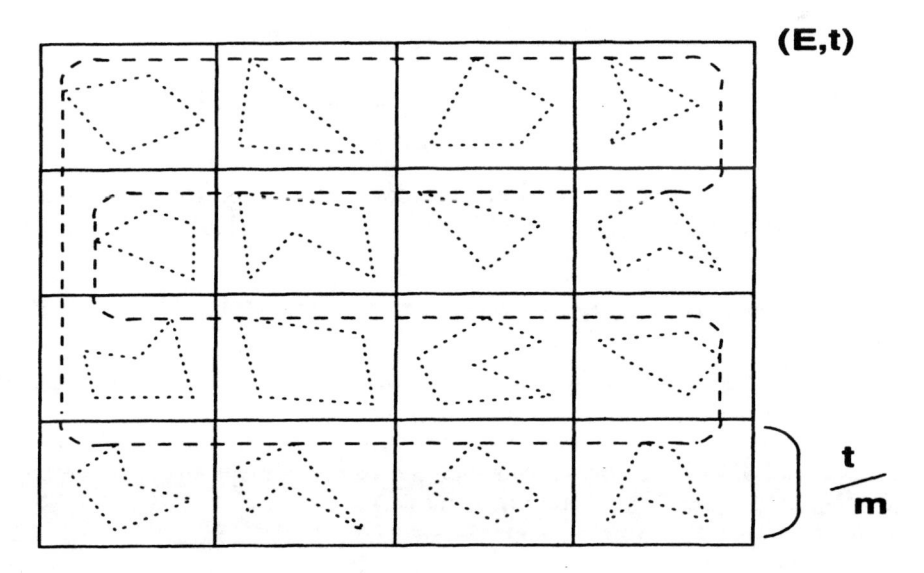

FIG. 2.1. *How to knit subsquare tours into a global tour.*

By applying (2.18) to the set $\Pi[0,t]^d$ and taking expectations, we find that in terms of $Z(t)$, we have a very simple recursion,

$$(2.19) \quad EZ(t) \leq m^d EZ(t/m) + Ctm^{d-1},$$

and it should be no surprise that we can extract the asymptotics of $EZ(t)$ from (2.19).

Replacing t by mt and dividing by $m^d t^d$ gives us

$$(2.20) \quad EZ(mt)/(m^d t^d) \leq EZ(t)/t^d + Ct^{1-d},$$

and if we take $t = 1$ in (2.20), we find

(2.21) $$0 \leq \gamma \equiv \liminf_{m \to \infty} EZ(m)/m^d \leq EZ(1) + C < \infty.$$

Now by the definition of γ for any $\epsilon > 0$, we can choose an integer t_0 as large as we like such that
$$EZ(t_0)/t_0^d + Ct_0^{1-d} \leq \gamma + \epsilon.$$

By the continuity of $Z(t)$, we can then find a δ such that for all $u \in (t_0, t_0 + \delta)$ we have

(2.22) $$EZ(u)/u^d + Cu^{1-d} \leq \gamma + 2\epsilon.$$

By (2.20), we therefore see that (2.22) also holds for all $u \in (mt_0, m(t_0 + \delta))$, and since for $m > t_0/\delta$ the intervals $((m+1)t_0, (m+1)(t_0 + \delta))$ and $(mt_0, m(t_0 + \delta))$ overlap, we have
$$[t_0^2/\delta, \infty) \subset \bigcup_{m=1}^{\infty} (mt_0, m(t_0 + \delta)),$$
so we see that (2.22) holds for all $u \geq t_0^2/\delta$. Hence we find
$$\limsup_{u \to \infty} EZ(u)/u^d \leq \liminf_{t \to \infty} EZ(t)/t^d + 2\epsilon,$$
so by the arbitrariness of $\epsilon > 0$, we arrive at the desired conclusion:

(2.23) $$EZ(t)/t^d \to \gamma \quad \text{as } t \to \infty.$$

Finally, we find that a change of variables in (2.23) and the representation (2.17) work together to give us basic asymptotic relationship for an average of the a_k's:

(2.24) $$\sum_{k=0}^{\infty} a_k e^{-t} t^k / k! \sim \gamma t^{(d-1)/d}.$$

The rest of the proof now follows a pattern that is particularly close to one used in the first chapter.

To exploit (2.24), we first need to show that a_k is reasonably smooth. We begin by noting that since we can join the tours through the first n points $\{X_1, X_2, \ldots, X_n\}$ and the second m points $\{X_{n+1}, X_{n+2}, \ldots, X_{n+m}\}$ at a cost of not more than $2\sqrt{d}$ to get a tour through $\{X_1, X_2, \ldots, X_{n+m}\}$, we can take expectations to see

(2.25) $$a_{n+m} \leq a_n + a_m + 2\sqrt{d}.$$

Now since we already know that $a_n = O(n^{(d-1)/d})$ and since (2.25) tells us that $0 \leq a_n - a_k \leq a_{n-k} + \sqrt{d}$ for $k \leq n$, we find the inequality that lets us see that sequence $\{a_n\}$ does not change rapidly:

(2.26) $$|a_n - a_k| \leq |a_{n-k} + 2\sqrt{d}| \leq c|n - k|^{(d-1)/d}.$$

If N is a Poisson random variable with mean n, we have

(2.27) $$Ea_N = \sum_{k=0}^{\infty} a_k e^{-n} n^k / k! \quad \text{and} \quad |a_n - Ea_N| \leq \sum_{k=0}^{\infty} |a_n - a_k| e^{-n} n^k / k!.$$

Since $\|X\|_1 \leq \|X\|_p$ for any random variable, we see from (2.27), the choice $p = 2d/(d-1) > 1$, and the fact that $\text{Var } N = n$ that $|a_n - Ea_N|$ is bounded by

$$c \sum_{k=0}^{\infty} |n-k|^{(d-1)/d} e^{-n} n^k/k! \leq c \left(\sum_{k=0}^{\infty} (n-k)^2 e^{-n} n^k/k! \right)^{(d-1)/2d} = cn^{(d-1)/2d}.$$

Thus we have proved that $|a_n - Ea_N| = O(n^{(d-1)/2d})$; so from the original asymptotic relation (2.24) and the bound (2.27), we find what we wanted:

(2.28) $\qquad EL_n/n^{(d-1)/d} \to \gamma \quad \text{as } n \to \infty.$

2.4. The Beardwood–Halton–Hammersley theorem.

We have now collected more than enough information to provide a proof of the central case of the remarkable theorem of Beardwood, Halton, and Hammersley (1959).

THEOREM 2.4.1 (BHH: uniform case). *If $\{X_i\}$ is a sequence of independent random variables with the uniform distribution on $[0,1]^d$, then there is a constant $0 < \beta_{\text{TSP}}(d) < \infty$ such that*

$$L(X_1, X_2, \ldots, X_n)/n^{(d-1)/d} \to \beta_{\text{TSP}}(d)$$

with probability one.

Proof. There is really nothing left to prove since we have already done all of the hard work. Still, to be polite, we should collect the evidence. The existence of $\beta_{\text{TSP}}(d)$ comes from (2.28), and the strict positivity of $\beta_{\text{TSP}}(d)$ comes from (2.16). Finally, the almost sure limit comes from Azuma's inequality as reflected in (2.11) and (2.12).

About the constants.

There is a natural curiosity to the constants of the BHH theorem, and considerable computation has been invested in the estimation of the $\beta_{\text{TSP}}(d)$. Because of the influence of scale—say, as reflected by the diameter of the unit d-cube—we place $\beta_{\text{TSP}}(d)$ on a more natural basis for comparison if we look at $\beta_{\text{TSP}}(d)/\sqrt{d}$. Beardwood, Halton, and Hammersley proved the following:

$$d = 2: \quad 0.44194 \leq \beta_{\text{TSP}}(2)/\sqrt{2} \leq 0.6508.$$
$$d = 3: \quad 0.37314 \leq \beta_{\text{TSP}}(3)/\sqrt{3} \leq 0.61771.$$
$$d = 4: \quad 0.34208 \leq \beta_{\text{TSP}}(4)/\sqrt{4} \leq 0.55696.$$

Beardwood, Halton, and Hammersley also proved that

(2.29) $\quad 0.24197 \leq \liminf \beta_{\text{TSP}}(d)/\sqrt{d} \leq \limsup \beta_{\text{TSP}}(d)/\sqrt{d} \leq 0.40825,$

and they naturally conjectured that one has a genuine limit in (2.29). This conjecture was established in Rhee (1992), where it is proved that one has the charming result

$$\beta_{\text{TSP}}(d)/\sqrt{d} \to \frac{1}{\sqrt{2\pi e}}.$$

Except for this limit case, we have very little analytical information on the constants $\beta_{\text{TSP}}(d)$. D. S. Johnson (personal communication) did extensive computer simulations using upper bounds based on the Lin–Kernighan heuristic and lower bounds based on the Held–Karp 1-tree relaxation to find with a very high degree of confidence that

$$0.70 \leq \beta_{\text{TSP}}(2) \leq 0.73,$$

but a formal proof of this result has not yet been given.

What did we use in the BHH theorem?

When we look back at the raw ingredients that went into the preceding proof, we find that only a few general properties of the TSP were used. The proof was designed with this intention and with more than a small dose of hindsight. A bit of scaling gave us (2.17) to start the ball rolling, and we got our basic push from (2.18). The next crucial bit was the smoothness of a_n expressed by (2.26), which provided the path to de-Poissonization. Even the finiteness and continuity of $EZ(t)$ can be extracted from (2.26). In later chapters, we will examine results that generalize what we have learned for the TSP, but before considering other problems or abstractions, we should nail down a few more developments for the TSP proper. In particular, we should show what happens for nonuniform random samples.

THEOREM 2.4.2 (Beardwood, Halton, Hammersley (1959)). *There is a constant $\beta_{\text{TSP}}(d)$ such that for any sequence of independent random variables $\{X_i\}$ with distribution μ with compact support, we have*

$$(2.30) \quad L(X_1, X_2, \ldots, X_n)/n^{(d-1)/d} \to \beta_{\text{TSP}}(d) \int_{\mathbb{R}^d} f(x)^{(d-1)/d}\, dx \quad \text{a.s.},$$

where $f(x)$ is the density of the absolutely continuous part of the distribution μ.

The first step of the proof is long enough to benefit from being considered independently as a lemma. What we need is a more precise expression of the fact that the length of the TSP on the $[0,1]^d$ is well approximated by the sum of the lengths of the shortest tours in a partition of $[0,1]^d$.

LEMMA 2.4.1. *There is a constant $c(d, m)$ depending only on d and m such that for any partition $\{Q_i\}$ of $[0,1]^d$ into m^d subcubes of side $1/m$, we have with probability one that*

$$(2.31) \quad L(X_1, X_2, \ldots, X_n) \leq \sum_{i=1}^{m^d} L(\{X_i : X_i \in Q_i\}) + c(d, m)$$

and

$$(2.32) \quad \sum_{i=1}^{m^d} L(\{X_i : X_i \in Q_i\}) \leq L(X_1, X_2, \ldots, X_n) + c(d, m)n^{(d-2)/(d-1)}.$$

Proof. The bound in (2.31) is easy. As we already noted in (2.18), we can tie the tours of the subcubes together to get a tour of the whole cube, and the cost of the connection is bounded by the cost of a tour through a set of representatives from each subcube.

The bound in (2.32) requires more thought. The idea is to take the tour T of the whole cube and build from it a tour of each of the subcubes. We let M be the set of points where T meets a face of one of the subcubes, and without loss of generality, we can suppose that T has no edge that meets a face except at a point and that no point of $\{X_1, X_2, \ldots, X_n\}$ is contained on the surface of any subcube. We will call these intesection points *marks*.

Now consider a typical subcube Q_i and suppose that there are m_i marks on the faces of Q_i. For each face $F_{s,i}$, $1 \leq s \leq 2d$, of Q_i that has an odd number of marks, we choose one mark, and we let O_i be the collection of these selected marks. To get a tour T_i of the points in Q_i, we first build $2d$ tours through the sets of unselected marks on the faces $F_{s,i}$ of Q_i. Since the number of unselected marks on a given face is even, there are an even number of edges in each of these tours. For a reason to be made clear in a moment, we double every other edge on each of these tours, and we will call the resulting graphs *reinforced tours*. What we angle toward with this fussing about with extra edges is that eventually we will want to use the elementary fact that any connected graph with only even degrees has an Eulerian tour (that is, an ordering of the adjacent edges so that all edges are used and none is used twice).

We now take these $2d$ reinforced tours together with the part of T that is contained in Q_i. This gives a graph G_i on $\{X_1, X_2, \ldots, X_n\} \cap Q_i$ and the unselected marks of $\cup F_{s,i}$ so that all of the degrees of each of the unselected marks is even; specifically, each such point is incident to three edges on the face from the reinforced tour and one edge that comes from the interior of the cube. Naturally, the degree of any point of $\{X_1, X_2, \ldots, X_n\} \cap Q_i$ is even since the only edges for such points are those of the tour T.

Now we choose one unselected point from each face that contains such a point, and we then add this point to $O = \cup_i O_i$. The cardinality of O is bounded independently of n, so we can make a T' tour of O at cost that is bounded independently of n. We now see that we can take T' together with the union of all the G_i's to form a connected graph that contains all of the marks and all the points of $\{X_1, X_2, \ldots, X_n\} \cap Q_i$. This graph has only vertices of even degree, so there is a tour of the graph that goes through all of the edges just once.[1] This tour naturally goes through all the points $\{X_1, X_2, \ldots, X_n\} \cap Q_i$, so its length provides an upper bound on the length of the shortened tour through $\{X_1, X_2, \ldots, X_n\} \cap Q_i$. Now we just need to estimate the total costs of all of these tours summed over all of the subcubes.

Since the faces of the Q_i have dimension $d-1$, the cost of G_i is bounded by $cm_i^{(d-2)/(d-1)}$, where c depends only on d and m. We therefore find

$$\sum_{i=1}^{m^d} L(\{X_i : X_i \in Q_i\}) \leq L(X_1, X_2, \ldots, X_n) + c \sum_{i=1}^{m^d} m_i^{(d-2)/(d-1)},$$

where c depends on d and m. Finally, since each edge in the original tour T can

---

[1] A less wasteful tour can be found by using a lemma of Karp (1977) that is discussed in the notes at the end of the chapter.

create a number K of marks that depends only on d and m, the sum of all of the m_i's is bounded by Kn, and the desired inequality (2.32) then follows by the application of Hölder's inequality.

We will now complete the proof of the Beardwood–Halton–Hammersley theorem by a sequence of steps from the uniform case to the general case.

Step 1: Densities with a special form. We will first prove the theorem for absolutely continuous random variables with density of the special form

$$f(x) = \sum_{k=1}^{m^d} a_k I_{Q_k}(x),$$

where $I_{Q_k}(\cdot)$ is the indicator function of the cube Q_k. The variables $\{X_i\}$ with this density can be thought of as being produced by an experiment where one chooses a cube Q_k with probability $a_k m^{-d}$ and then chooses a point at random from Q_k. By using the uniform version of the BBH theorem (adjusted for scale) and by applying the strong law of large numbers for the Bernoulli random variables $I_{Q_k}(X_i)$, we see without difficulty that we have with probability one that

$$L(\{X_i : 1 \leq i \leq n,\ X_i \in Q_k\}) \sim \frac{1}{m}(a_k m^{-d} n)^{(d-1)/d}.$$

When we sum this relation over $1 \leq k \leq m^d$ and use (2.31) and (2.32) to relate the subsquare tours to the whole tour, we get exactly (2.30) for our special densities.

Step 2: General densities. Now consider the more general case of absolutely continuous random variables with density f that has compact support in the $[0,1]^d$'s and for which the integral in (2.30) is finite. For any such f and any $\epsilon > 0$, there is a density ϕ of the special form just used such that

(2.33) $\quad \int_{\mathbb{R}^d} |\phi(x) - f(x)|\, dx \leq \epsilon \quad$ and $\quad \int_{\mathbb{R}^d} |\phi(x) - f(x)|^{(d-1)/d}\, dx \leq \epsilon.$

By a standard coupling argument,[2] the first condition of (2.33) is enough to guarantee that there is a joint distribution for random variables (X, Y) such that X has density f, Y has density ϕ, and $P(X \neq Y) \leq 2\epsilon$. We now let $Z_i = (X_i, Y_i)$ be independent random variables with the joint distribution given by the coupling distribution. We will extract all we need from the natural chain of inequalities

$L(\{X_i : 1 \leq i \leq n,\ X_i = Y_i\})$

$\qquad \leq L(X_1, X_2, \ldots, X_n)$

$\qquad \leq L(\{X_i : 1 \leq i \leq n,\ X_i = Y_i\}) + L(\{X_i : 1 \leq i \leq n,\ X_i \neq Y_i\}) + 2\sqrt{d}$

$\qquad \leq L(Y_1, Y_2, \ldots, Y_n) + c(\text{card}\{1 \leq i \leq n,\ : X_i \neq Y_i\})^{(d-1)/d} + 2\sqrt{d}.$

(2.34)

---

[2]For example, the γ coupling of Lindvall (1992, p. 18).

We now write $L_n = L(X_1, X_2, \ldots, X_n)$, divide (2.34) by $n^{(d-1)/d}$, and let $n \to \infty$ to find from the first and last elements of the chain that

$$(2.35) \quad (1-2\epsilon)^{(d-1)/d} \int_{\mathbb{R}^d} \phi^{(d-1)/d}(x)\,dx \leq \liminf_{n\to\infty} L_n/n^{(d-1)/d}$$

$$\leq \limsup_{n\to\infty} L_n/n^{(d-1)/d} \leq \int_{\mathbb{R}^d} \phi^{(d-1)/d}(x)\,dx + (2\epsilon)^{(d-1)/d}.$$

By the second inequality of (2.33), we can replace ϕ with f in (2.35) at a cost that is measured in ϵ's, and then by the arbitrariness of ϵ, we see that the proof of this step is complete.

Step 3: Purely singular distributions. We now consider the case where the distribution μ is purely singular with compact support in $[0,1]^d$. Since the support of μ is contained in a set of Lebesgue measure zero, for any $\epsilon > 0$ we can choose an m and a partition of $[0,1]^d$ into m^d cubes Q_k such that we have a subset $S \subset [1, m^d]$ with

$$\mu\left\{\bigcup_{k \in S} Q_k\right\} \geq 1 - \epsilon \quad \text{with card } S \leq \epsilon m^d.$$

We then write the sample as the union of the points in a cube of S or in the complement of the set of such cubes; we have

$$L_n \leq cm^{(d-1)/d} + \sum_{k \in S} L(X_i : X_i \in Q_k) + L\left(X_i : X_i \notin \bigcup_{k \in S} Q_k\right).$$

On taking the limit as $n \to \infty$ in the last inequality, we have

$$(2.36) \quad \limsup_{n\to\infty} L_n/n^{(d-1)/d} \leq \frac{1}{m} \sum_{k \in S} \mu(Q_k)^{(d-1)/d} + c\epsilon^{(d-1)/d}.$$

Now since the sum of the $\mu(Q_i)$ is bounded by one, we find from Hölder's inequality and the definition of S that we have

$$\frac{1}{m} \sum_{k \in S} \mu(Q_k)^{(d-1)/d} \leq \frac{1}{m} (\operatorname{card} S)^{1/d} \leq \epsilon^{1/d}.$$

When we return this last bound to (2.36), we see that the purely singular case of the theorem is complete.

Step 4: The general case. For the final step of the proof, one needs to consider the case of X_i with a distribution with compact support and both singular and absolutely continuous parts. If F denotes the support of the singular part and F^c denotes the complement of F, then the final step of the proof is completed by use of the decomposition

$$L(X_i : X_i \in F^c) \leq L_n \leq L(X_i : X_i \in F^c) + L(X_i : X_i \in F) + 2\sqrt{d},$$

where the limit results that have already been obtained for the purely singular and purely absolutely continuous cases now suffice to carry the day. The proof of Theorem 2.4.2 is thus complete.

One of the issues that was left unresolved by Beardwood, Halton, and Hammersley (1959) was treatment of the unbounded case, though a conjecture was offered for a necessary and sufficient on f that would guarantee (2.30). This conjecture was only recently resolved by Rhee (1993b); and, although the tidy condition suggested by Beardwood, Halton, and Hammersley (1959) turns out not to be strong enough, Rhee (1993b) provides a sufficient condition that speaks clearly to the original intention—though the condition is a bit more technical. We first consider Rhee's sufficient condition, and then we will relate it to simpler ones.

THEOREM 2.4.3 (Rhee (1993b)). *Suppose that the random variables $\{X_i\}$ are independent with absolutely continuous distribution and density $f(x)$. We let A_k denote the annular shell in $\mathbb{R}^d$ with inner radius 2^k and outer radius 2^{k+1}, and we set*

$$a_k(f) = 2^{kd/(d-1)} \int_{A_k} f(x)\,dx.$$

If the density f satisfies

(2.37) $$\sum_{k\geq 1} a_k(f)^{(d-1)/d} < \infty,$$

then

$$L(X_1, X_2, \ldots, X_n)/n^{(d-1)/d} \to \beta_{\text{TSP}}(d) \int_{\mathbb{R}^d} f(x)^{(d-1)/d}\,dx \quad a.s.$$

Before discussing the proof of this result, we should first note that the sufficient condition (2.37) is the best that one can do in terms of conditions on $a_k(f)$. In particular, Rhee (1993b) also showed that for any real sequence $\{a_k\}$ such that $a_k > 0$ and

$$\sum_{k\geq 1} a_k^{(d-1)/d} = \infty,$$

there is a density f on $\mathbb{R}^d$ such that $a_k(f) \leq a_k$, yet for which

(2.38) $$\int_{\mathbb{R}^d} f(x)^{(d-1)/d}\,dx < \infty \quad \text{and} \quad \lim_{n\to\infty} L_n/n^{(d-1)/d} = \infty \quad a.s.$$

To relate Rhee's condition (2.37) to the more traditional moment bounds, we first note that

(2.39) $$\int_{\mathbb{R}^d} \|x\|^{d/(d-1)} f(x)\,dx < \infty \quad \Leftrightarrow \quad \sum_{k\geq 1} a_k(f) < \infty,$$

so Rhee's condition is certainly stronger than the conditions of (2.39). Moreover, by Rhee's examples reflected in (2.38), we see that the moment condition in (2.39) is not strong enough to yield a limit theorem. Still, the moment condition (2.39) is not too far away from Rhee's sufficient condition since by an easy application of Hölder's inequality, one can show that

$$\int_{\mathbb{R}^d} \|x\|^\gamma f(x)\, dx < \infty \quad \text{for } \gamma > d/(d-1) \quad \Rightarrow \quad \sum_{k \geq 1} a_k(f)^{(d-1)/d} < \infty.$$

We will not give a full proof of Theorem 2.4.3, but we will show the how one uses the geometry of the TSP to reduce the problem to one in pure analysis. We first note that for *any density*, we have one side of the problem well under control since

$$\liminf_{n \to \infty} L_n/n^{(d-1)/d} \geq \beta_{\text{TSP}}(d) \int_{\mathbb{R}^d} f(x)^{(d-1)/d}\, dx \quad \text{a.s.}$$

Thus the whole issue turns on the development of a bound on the limit supremum.

To work toward such a bound, we take $1 < q < \infty$ and let $L_{n,q}^*$ be the length of the shortest tour through $\{X_i : 1 \leq i \leq n, \|X_i\| \leq 2^q\}$. Similarly, let $L_{n,k}$ be the length of the shortest tour of the points in the set $\{X_i : 1 \leq i \leq n, 2^k \leq \|X_i\| \leq 2^{k+1}\}$, the cardinality of which we denote as $N(n,k)$. We also let $s(n)$ denote the largest k such that $N(n,k) \neq 0$.

We need just a small variation on our familiar arguments to derive a natural upper bound:

$$L_n \leq L_{n,q}^* + \sum_{k:k \geq q} L_{n,k} + 2^{s(n)+4}.$$

The variation we require concerns the last term and reflects the way that the subtours are tied together. If S denotes set of points consisting of one representative from each of the subtours, then we construct a tour through the points of S by first visiting our points in increasing order of their distance from the origin and then returning to the first representative. By crude bounds on the lengths of the successive edges and by summing a geometric series, one gets $2^{s(n)+4}$ or a little better as the cost of the tour through S.

From the BHH theorem for distributions with compact support, we have for any fixed integer $q \geq 1$ that

$$\limsup_{n \to \infty} L_{n,q}^*/n^{(d-1)/d} \leq \beta_{\text{TSP}}(d) \int_{\mathbb{R}^d} f(x)^{(d-1)/d}\, dx \quad \text{a.s.},$$

and by our L^∞ bound, we have

$$L_{n,k} \leq c 2^k N(n,k)^{(d-1)/d},$$

so the whole issue comes down to showing that with probability one we have

$$\limsup_{n \to \infty} \sum_{k:k \geq q} \left(2^k N(n,k)^{(d-1)/d} + 2^{s(n)} \right) \leq h(q),$$

where $h(q)$ is a real function for which we can show that $h(q) \to 0$ as $q \to \infty$. At this point, the geometry of the TSP has escaped from the picture, and one is left with a question in hard analysis for which we do best to refer to Rhee (1993b).

2.5. Karp's partitioning algorithms.

In our analysis of the probability theory of the TSP, we found that partitioning the square into subsquares led us to useful analytical relations. The same insight is relevant to the construction of many algorithms that apply to problems of Euclidean combinatorial optimization.

In particular, Karp (1976) observed that under a variety of natural probability models, one can build fast algorithms that will yield nearly optimal solutions with probability one. This discovery was especially engaging in the case of the TSP, where Papadimitrou (1978a) had proved that even in the case of the Euclidean plane, the problem of determining a shortest path is NP-complete. Thus Karp's partitioning algorithm gave the first example of an NP-complete problem for which there exists a polynomial-time algorithm such that with probability one the algorithm provides a solution that is within a factor of $1 + \epsilon$ times the value of the optimal solution.

In Karp (1977), the idea of the partitioning heuristic was analyzed in detail for the TSP in $d = 2$. The essence of the idea is that one chooses a partition based on the problem size in such a way that the problem that one has to solve in each of the subsquares can be done in time that is polynomial in n even if it has to be exponential in the cardinality of the number of points in the subsquare. For example, the dynamic-programming algorithm of Bellman (1962) will solve an n-point traveling-salesman problem in time $n^2 2^n$, so if we arrange our affairs so that with sufficiently high probability no subsquare has more than $\log n$ points, then the whole set of $O(n/\log n)$ optimal subtours can be computed in time $O(n^4)$. The cost of tying these subtours together is provably small compared to the sum of the lengths of the subtours, and the time to tie the subtours together is smaller than the time to compute the optimal subtours.

Perhaps this description is too quick and too dirty, but still the point should be clear. If one proceeds with care, much better can be done. Nowadays, one natural phrasing of the main results of Karp (1977) can be put together as the following theorem.

THEOREM 2.5.1 (Karp (1977)). *Suppose that* $\{X_1, X_2, \ldots, X_n, \ldots\}$ *are independent random variables with the uniform distribution on* $[0, 1]^2$. *For every* $\epsilon > 0$ *and* $n \geq 1$, *there is a partitioning algorithm* $\mathcal{A}(\epsilon, n)$ *such that* $\mathcal{A}(\epsilon, n)$ *runs to completion in time that is always bounded by* $\mathcal{C}(\epsilon) n + O(n \log n)$ *and the tour of* $\{X_1, X_2, \ldots, X_n\}$ *that is produced* $\mathcal{A}(\epsilon)$ *by has length* K_n *that satisfies*

$$(2.40) \qquad \sum_{i=1}^{\infty} P(K_n > L_n(1 + \epsilon)) < \infty.$$

The informal way of stating this result is to say that $\mathcal{A}(\epsilon)$ produces a path that with probability one is ϵ-*optimal*. The distinction that sometimes needs to be borne in mind in such statements is that there are two natural ways to look at a sequence of problems of size n. The distinction turns on the nature of the dependence or independence of the problem that follows the problem based on $\{X_1, X_2, \ldots, X_n\}$.

For example, one can have the problem based on $\{X_1, X_2, \ldots, X_{n+1}\}$—this is the *incrementing model of problem generation*—or one can have a problem based on a whole fresh sample $\{X'_1, X'_2, \ldots, X'_{n+1}\}$—the *independent model of problem generation*. This distinction is simple enough from the probabilistic perspective, but until the point was made clearly by Weide (1978), the algorithmic waters were often muddy.

With the type of complete convergence expressed in (2.40), the Borel–Cantelli lemma lets us simultaneously accommodate the more stringent independent model as well as the incrementing model. An easy proof of (2.40) is given in Karp and Steele (1990), but with the tools that have been developed earlier in this chapter, one should find no difficulty in supplying the proof of even stronger assertions.

Naturally, the partition heuristic can be applied in d-dimensions. Moreover, the running-time rate result $\mathcal{C}(\epsilon)n + O(n \log n)$ can be improved, though there is no known algorithm that runs in linear time for fixed $\epsilon > 0$. For details on both the d-dimensional result and the speedup to near linear running time, one should consult Halton and Terada (1982).

2.6. Introduction to the space-filling curve heuristic.

Today, every baby knows[3] that there are continuous maps from the unit interval $[0, 1]$ onto the unit square $[0, 1]^2$. This fact once defied the intuition of the masters and even stirred concern about the meaning of dimension, but now such curves are often regarded as humdrum and little-used curiosities. This view evolved naturally, but the time has come for revisionary thinking. Space-filling curves are not only quaint; they are also useful. Here we will make several uses of such curves—particularly analytical uses—but we will first consider how they provide a simple heuristic for the TSP.

We should note that when we are concerned with computation, not just any old space-filling curve will do. To be useful computationally, our surjective mapping $\psi : [0,1] \to [0,1]^2$ needs the property that for each $x \in [0,1]^2$ one can *quickly compute* some $t \in [0,1]$ such that $\psi(t) = x$. The set of such ψ's turns out to be large. A technical point that we should note here is that we cannot speak of computing the inverse of ψ since our ψ *cannot* be a one-to-one mapping by basic topological considerations.

The natural hope implicit in the space-filling curve heuristic is that we may be able to find a reasonably short tour through a set of n points $\{X_1, X_2, \ldots, X_n\} \subset [0,1]^2$ by visiting the points in the order of their preimages in $[0,1]$. Formally, we have a three-step process:

- We compute a set of points $\{t_1, t_2, \ldots, t_n\} \subset [0,1]$ such that $\psi(t_i) = x_i$ for each $1 \leq i \leq n$,

---

[3] A term of art regularly used by P. Erdös in lectures when introducing a fact, either well known or, sometimes, not so well known.

42   CHAPTER 2

- we order the t_i so that $t_{(1)} \leq t_{(2)} \leq \ldots \leq t_{(n)}$, and
- we define a permutation $\sigma : [1, n] \to [1, n]$ by requiring that $x_{\sigma(i)} = \psi(t_{(i)})$.

The path that visits $\{X_1, X_2, \ldots, X_n\}$ in the order of $x_{\sigma(1)}, x_{\sigma(2)}, \ldots, x_{\sigma(n)}$ will be called the *space-filling curve path*, and the tour that closes this path by adding the step from $x_{\sigma(n)}$ back to $x_{\sigma(1)}$ will be called the *space-filling curve tour* (see Figure 2.2.).

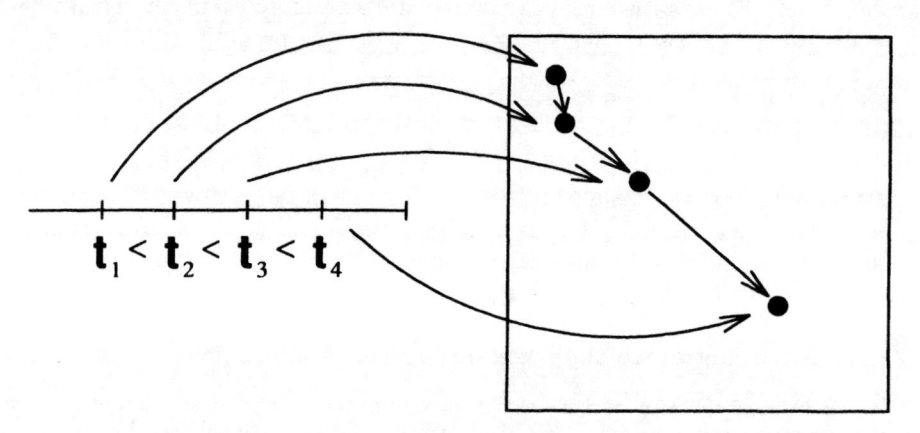

FIG. 2.2. *Space-filling curve heuristic.*

For the space-filling curve ψ heuristic to be effective, one wants ψ to be as smooth as possible since the smoothness of ψ is closely related to the shortness of the tour one obtains. Many of the classical space-filling curves are Lipschitz of order $\frac{1}{2}$, which is to say there exists a constant c_ψ so that for any $0 \leq s, t \leq 1$ one has $\|\psi(s) - \psi(t)\| \leq c_\psi |s - t|^{\frac{1}{2}}$. One can easily check that no space-filling curve can be Lipschitz of order greater than $\frac{1}{2}$ since for all k the union of the images of $[i/k, (i+1)/k]$ for $0 \leq i < k$ must cover $[0,1]^2$, and as a consequence, one or more of these images must have diameter at least $2\pi^{-\frac{1}{2}} k^{-\frac{1}{2}}$.

The existence of Lipschitz-$\frac{1}{2}$ space-filling curves gives us an analytical tool that often makes easy work of analytical facts that might otherwise require considerable ingenuity. For example, here is an analytical assertion that makes an excellent exercise:

There is a constant c such that for any $\{x_1, x_2, \ldots, x_n\} \subset [0,1]^2$ there is a permutation σ of $[n]$ such that

$$(2.41) \qquad \sum_{i=1}^{n-1} |x_{\sigma(i)} - x_{\sigma(i+1)}|^2 < c.$$

This charming fact does not yield to kneejerk resolution. If one tries the obvious idea of using the fact that any n points in the square must contain a pair within a distance of $O(1/\sqrt{n})$, then without coming up with other ideas, we only get a bound of order $O(\log n)$—not the $O(1)$ bound required in (2.41). In contrast, the space-filling curve heuristic does the job neatly.

Following the three-step process, there are $0 \leq t_{(1)} \leq t_{(2)} \leq \cdots \leq t_{(n)} \leq 1$ such that
$$\sum_{i=1}^{n-1} |x_{\sigma(i)} - x_{\sigma(i+1)}|^2 = \sum_{i=1}^{n-1} |\psi(t_i) - \psi(t_{i+1})|^2,$$
and this representation makes the proof of the fact automatic. We just apply the Lipschitz condition and then take advantage of telescoping sums:

(2.42) $$\sum_{i=1}^{n-1} |\psi(t_i) - \psi(t_{i+1})|^2 \leq c_\psi^2 \sum_{i=1}^{n-1} |t_i - t_{i+1}| \leq c_\psi^2.$$

For the moment, one may regard (2.42) as something of a curiosity, but we will see in Chapter 6 that (2.42) helps us make light work of some analytical problems that had earlier required extensive, delicate work.

With the pattern of proof of (2.42) in mind, many facts about points in the unit square become evident when otherwise the proof of the such facts might have required considerable cleverness with pigeonhole arguments or inductions. As an added bonus, the appearance of the Lipschitz constant c_ψ in relations like (2.42) can give extra meaning to the constants that appear in our geometric bounds. Finally, we should also note that by (2.41) and Schwarz's inequality, we find the L_∞ bound for $d = 2$ that we have often used before, $\|L_n\|_\infty \leq c_\psi \sqrt{n}$, though this time there is an additional interpretation of the constant.

The analytical use of space-filling curves is not restricted to worst-case L^∞ bounds, nor is it restricted to $d = 2$. For all $d \geq 2$, there are continuous functions from $[0, 1]$ onto $[0, 1]^d$ that are Lipschitz of order $1/d$ and which have a further property that makes them useful in probabilistic investigations. The additional property is that they are measure preserving in the sense that for any Borel set $A \subset [0,1]^d$, $\lambda_1(\psi^{-1}(A)) = \lambda_d(A)$, where λ_d denotes the Lebesgue measure on $\mathbb{R}^d$. This property helps translate many questions about random samples in $[0, 1]^d$ to simpler questions about a random sample in $[0, 1]$. The recipe is simple; one takes U_i, $1 \leq i \leq n$, independent with the uniform distribution on $[0, 1]$, and notes that by the measure-preserving property, if we define $X_i \equiv \psi(U_i)$, then the X_i's are independent with the uniform distribution on $[0, 1]^d$. One then exploits the linear structure of $[0, 1]$ and the theory of order statistics of $[0, 1]$ to resolve the question at hand.

2.7. Asymptotics for the space-filling curve heuristic.

There is a challenge that is implicit with any heuristic for the TSP. To what extent does one have the same probabilistic behavior for the heuristic that one has for the optimal TSP tour length? Sometimes the challenge can be resolved by general theorems of a type that we will develop shortly, but sometimes the basic underlying behavior is just different. This exceptional—but intriguing— situation is exactly what we have in the case of the space-filling curve heuristic. The basic discovery of this phenomenon is due to Platzman and Bartholdi (1989), but we will follow the development from a sightly more general perspective that was developed in Gao and Steele (1994).

A general result.

So far, all we required of our space-filling curves is that they be measure preserving and Lipschitz of order $\frac{1}{2}$, but to get a result that comes closer to the Beardwood–Halton–Hammersley theorem, we will need to know more about our curve ψ. Many of the classical space-filling curves, such as those of Hilbert (1891) or Peano (1890), have important aspects of self-similarity, and self-similarity is just the tool we need to propel a more searching asymptotic analysis. Specifically, we focus on mappings that have the following properties:

A1: *Dilation property.* There is an integer $p \geq 2$ such that for all $0 \leq s, t \leq 1$,
$$\|\psi(s) - \psi(t)\| = \sqrt{p} \left\|\psi\left(\frac{s}{p}\right) - \psi\left(\frac{t}{p}\right)\right\|.$$

A2: *Translation property.* For $1 \leq i \leq p$, if $(i-1)/p \leq s, t \leq i/p$, then
$$\|\psi(s) - \psi(t)\| = \|\psi(s + 1/p) - \psi(t + 1/p)\|.$$

A3: *Bi-measure-preserving property.* Given any Borel set A in $[0, 1]$, one has
$$\lambda_1(A) = \lambda_2(\psi(A)),$$
where λ_d is the Lebesgue measure on $\mathbb{R}^d$.

Pleasantly, just these properties turn out to be sufficient to determine the asymptotics of the expected value of the length L_n^{SFC} of the tour given by the space-filling curve heuristic based on ψ.

THEOREM 2.7.1. *If a heuristic tour is built using a space-filling curve ψ that satisfies the dilation property (A1), the translation property (A2), and the bi-measure-preserving property (A3), then there exists a continuous function φ of period one such that*
$$\lim_{n \to \infty} \frac{EL_n^{\text{SFC}}}{\sqrt{n\varphi(\log_p n)}} = 1 \quad a.s.,$$
where p is the integer appearing in assumptions A1 and A2.

The punchline of this result is that one comes close to the behavior of the BHH theorem, but there remains a final difference that never goes away. The periodic function in the denominator will always produce a kind of oscillatory behavior that is not present in the behavior of the optimal tour length. One point that is intriguing from the point of view of simulation analysis is that the logarithmic term guarantees that the periods between successive maxima and minima of $\varphi(\log_p n)$ are longer and longer, so one might have been easily convinced by simulations that one has convergence of the naïve ratio $EL_n^{\text{SFC}}/\sqrt{n}$. Yet we know that $\|\varphi\|_\infty > 0$, so this ratio definitely does not converge.

To be able to extract strong limit laws for L_n^{SFC} from the knowledge of the asymptotics of the expected value, we naturally need some information about the concentration of L_n^{SFC} about its mean. The next result provides more than enough information in the situations of principle concern. The proof of the concentration inequality will also be of interest to us because it provides a nice illustration of how some central results of martingale theory can be brought into play.

THEOREM 2.7.2. *If a space-filling curve φ has the bi-measure-preserving property (A3), then there are constants A and B such that for all $t \geq 0$,*

$$P(|L_n^{\text{SFC}} - EL_n^{\text{SFC}}| \geq t) \leq B\exp(-At^2/\log t).$$

We will not prove Theorem 2.7.1 here since the proof is long and analytical, but we will prove Theorem 2.7.2 because of the lessons that emerge concerning the role of martingales. The first lemma is pure martingale theory, though there are combinatorial features to the proof. The proof is also quite general and can be applied almost without change to the optimal TSP tour length.

LEMMA 2.7.1. *Let d_i, $1 \leq i \leq n$, be a martingale-difference sequence. If there are two constants c_1 and c_2 such that for $p > 1$ and $1 \leq i \leq n$ we have*

$$\|d_i\|_\infty \leq c_1(n-i+1)^{-1/2}$$

and

(2.43) $$\|d_i\|_p \leq c_2(p/n)^{1/2},$$

then there is a constant c_3 not depending on p or n such that

(2.44) $$\left\|\sum_{i=1}^n d_i\right\|_p \leq c_3 p^{1/2}(\log p)^{1/2}.$$

Proof. An important ingredient of the proof is the *square function* associated to a martingale difference sequence $\{\xi_i\}$ by

$$S_n(\xi) = \left(\sum_{i=1}^n \xi_i^2\right)^{1/2}.$$

The square function is often more tractable than the original martingale, and by a remarkable theorem of Burkholder (1966, 1973), the L_p norm of $S_n(\xi)$ is still comparable with the L_p norm of the original martingale. Specifically, for $1 < p < \infty$ and $1/p + 1/q = 1$, we have

$$\left\|\sum_{i=1}^n \xi_i\right\| \leq 18pq^{1/2}\|S_n(\xi)\|_p.$$

To use Burkholder's inequality, we will first check that the martingale differences $\{d_i\}$ of the lemma satisfy the more specialized inequality. To start the ball rolling, we take an arbitrary $S \subset [n]$ and examine an integer power of the squares of the $\{d_i\}$ over S. By direct expansion and the generalized Hölder inequality, we find

$$E\left(\sum_{i \in S} d_i^2\right)^p = \sum_{i_1 \in S}\sum_{i_2 \in S}\cdots\sum_{i_p \in S} E d_{i_1}^2 d_{i_2}^2 \cdots d_{i_p}^2$$

$$\leq \sum_{i_1 \in S}\sum_{i_2 \in S}\cdots\sum_{i_p \in S} E(d_{i_1}^{2p})^{1/p} E(d_{i_2}^{2p})^{1/p} \cdots E(d_{i_p}^{2p})^{1/p}.$$

Since by our hypothesis (2.43) we have $Ed_i^{2p} \leq c_2^{2p}(2p/n)^p$, we can apply this bound in the last expression to give

$$\left\|\left(\sum_{i \in S} d_i^2\right)^{1/2}\right\|_{2p}^{2p} = E\left(\sum_{i \in S} d_i^2\right)^p \leq (\text{card } S)^p c_2^{2p}(2p/n)^p,$$

or in the form in which we will use the bound,

(2.45) $$\left\|\left(\sum_{i \in S} d_i^2\right)^{1/2}\right\|_{2p} \leq (\text{card } S)^{1/2} c_2 (2p/n)^{1/2}.$$

The last bound provides just what we need in Burkholder's inequality. To put this to work toward (2.44), we first write

(2.46) $$\left\|\sum_{i=1}^n d_i\right\|_{2p} \leq \left\|\sum_{i \leq \alpha n} d_i\right\|_{2p} + \left\|\sum_{i > \alpha n} d_i\right\|_{2p}$$

and note that the second term is estimated by (2.45) and Burkholder's inequality to give

$$\left\|\sum_{i > \alpha n} d_i\right\|_{2p} \leq 18(2p)\left(\frac{2p}{2p-1}\right)^{1/2} c_2(2p)^{1/2}(1-\alpha)^{1/2}.$$

The first term of the right-hand side of (2.46) can also be estimated easily since by integrating the bound that we get from Azuma's inequality, we have

$$E\left|\sum_{i \leq \alpha n} d_i\right|^p \leq p \int_0^\infty t^{p-1} 2 \exp\left(\frac{-t^2}{2}\left\{\sum_{i \leq \alpha n} \|d_i\|_\infty\right\}^{-1}\right) dt,$$

or, on applying the crudest Γ-function bound,

$$\left\|\sum_{i \leq \alpha n} d_i\right\|_p \leq c_4 p^{1/2}\left(\sum_{i \leq \alpha n}(n-i+1)^{-1}\right)^{1/2}.$$

The harmonic sum is easily estimated and gives us what we need to bound the first of the two terms on the right side of (2.46),

$$\left\|\sum_{i \leq \alpha n} d_i\right\|_{2p} \leq c_5 p^{1/2}\left(\log \frac{1}{\alpha}\right)^{1/2}.$$

We finally choose α to minimize the sum of the bounds on the two terms that make up (2.46). The choice $(1-\alpha)^{1/2} = p^{-1}$ together with a small amount of arithmetic will complete the proof of the lemma.

CONCENTRATION OF MEASURE AND THE CLASSICAL THEOREMS 47

The clearest way to exploit L^p bounds like those just obtained is to note their equivalence to tail bounds. This is an important principle, but once one has the idea, the implementation is easy—just use Markov's inequality for a wise choice of p.

LEMMA 2.7.2. *For any random variable Z, there are constants a and b such that*

$$P(|Z| \geq t) \leq a \exp(-bt^2/\log t), \quad t \geq 0,$$

if and only if there is a constant c such that for all $p \geq 1$,

$$\|Z\|_p \leq c p^{1/2} (\log p)^{1/2}.$$

We can now close the loop on the problem that motivated our excursion into L^p, the derivation of a near sub-Gaussian tail bound for the length of the tour provided by the space-filling curve heuristic. The proof of Theorem 2.7.1 again calls on the familiar martingale differences,

$$(2.47) \qquad d_i = E(L_n^{\text{SFC}}|\mathcal{F}_i) - E(L_n^{\text{SFC}}|\mathcal{F}_{i-1}),$$

and all of the work goes into showing that these martingale differences satisfy the conditions of the lemma.

To reexpress the conditional expectations of (2.47) in a more convenient form, we introduce independent random variables $\{\tilde{X}_i : 1 \leq i \leq n\}$ with the uniform distribution on $[0,1]^2$ that are also independent of $\{X_i : 1 \leq i \leq n\}$, and we define $L_n^{\text{SFC}}(i) = L^{\text{SFC}}(X_1, X_2, \ldots, X_{i-1}, \tilde{X}_i, X_{i+1}, \ldots, X_n)$. The purpose of introducing these variables is that we find the new representation

$$d_i = E(L_n^{\text{SFC}} - L_n^{\text{SFC}}(i)|\mathcal{F}_i).$$

Now we need to find bounds on d_i. Since we assume that ψ is a measure-preserving mapping, we can assume that our independent random variables $\{X_i\}$ and $\{\tilde{X}_i\}$ are represented as $\psi(t_i) = X_i$ and $\psi(\tilde{t}_i) = \tilde{X}_i$ for all $1 \leq i \leq n$ where the sequences $\{t_i\}$ and $\{\tilde{t}_i\}$ are independent with the uniform distribution on $[0,1]$.

We need to obtain some information about the sizes of the gaps of the order statistics of a random sample from the unit interval. For any $x, y_1, y_2, \ldots, y_k \in [0,1]$, we think of $G(x; y_1, y_2, \ldots, y_k)$ as the size of the interval of the order statistics of $\{y_1, y_2, \ldots, y_k\}$ into which x falls, but to make the appropriate connection to the space-filling curve heuristic, we need to make a less conventional choice at the ends of the interval. Formally, if $0 < y_{(1)} < y_{(2)} < \cdots < y_{(k)}$, we define

$$G(x; y_1, y_2, \ldots, y_k) = \begin{cases} |y_{(i)} - y_{(i+1)}| & \text{if } x \in (y_{(i)}, y_{(i+1)}], \\ 1 + |1 - y_{(n)}| & \text{if } x \in (y_{(n)}, 1], \\ 1 + y_{(1)} & \text{if } x \in [0, y_{(1)}]. \end{cases}$$

If we use the hat notation to indicate an element that is missing from a set (so $\{s_1, s_2, \hat{s}_3\} = \{s_1, s_2\}$), then we have by the definition of the space-filling heuristic and the Lipschitz condition on ψ that

$$\begin{aligned}0 &\leq L^{\mathrm{SFC}}(X_1, X_2, \ldots, X_i, \ldots, X_n) - L^{\mathrm{SFC}}(X_1, X_2, \ldots, \hat{X}_i, \ldots, X_n) \\ &\leq 2c_\psi G^{1/2}(X_i; X_1, X_2, \ldots, \hat{X}_i, \ldots, X_n) \\ &\leq 2c_\psi G^{1/2}(X_i; X_{i+1}, \ldots, X_n),\end{aligned}$$

and in a perfectly parallel way,

$$\begin{aligned}0 &\leq L^{\mathrm{SFC}}(X_1, X_2, \ldots, \tilde{X}_i, \ldots, X_n) - L^{\mathrm{SFC}}(X_1, X_2, \ldots, \hat{X}_i, \ldots, X_n) \\ &\leq 2c_\psi G^{1/2}(\tilde{X}_i; X_1, X_2, \ldots, \hat{X}_i, \ldots, X_n) \\ &\leq 2c_\psi G^{1/2}(\tilde{X}_i; X_{i+1}, \ldots, X_n),\end{aligned}$$

so we find

$$|L_n^{\mathrm{SFC}} - L_n^{\mathrm{SFC}}(i)| \leq 2c_\psi G^{1/2}(X_i; X_{i+1}, \ldots, X_n) + 2c_\psi G^{1/2}(\tilde{X}_i; X_{i+1}, \ldots, X_n).$$

The reason for restricting the set of variables in our bound down to $X_{i+1}, \ldots, X_n$ is brought out when we take expectations and use independence to get

$$|d_i| = |E(L_n^{\mathrm{SFC}} - L_n^{\mathrm{SFC}}(i) \mid \mathcal{F}_i)| \leq E(|L_n^{\mathrm{SFC}} - L_n^{\mathrm{SFC}}(i)| \mid \mathcal{F}_i)$$

$$\leq 2c_\psi E(G^{1/2}(X_i; X_{i+1}, \ldots, X_n) \mid X_i) + 2c_\psi EG^{1/2}(\tilde{X}_i; X_{i+1}, \ldots, X_n).$$
(2.48)

But now, by the same type of calculation we used in Lemma 2.1.1, we find that for all $x \in [0,1]^2$ we have

$$EG^{1/2}(x; X_{i+1}, \ldots, X_n) \leq c(n - i + 1)^{-1/2},$$

and by applying this bound in (2.48), we see that we have established the first condition of the lemma.

All that remains for the proof of Theorem 2.7.2 is to show that there is constant c such that

$$\|d_i\|_p \leq c(p/n)^{1/2}.$$

By the definition of d_i and Jensen's inequality, we have

$$E|d_i|^p = E|E(L_n^{\mathrm{SFC}} - L_n^{\mathrm{SFC}}(i)|\mathcal{F}_i)|^p \leq E|L_n^{\mathrm{SFC}} - L_n^{\mathrm{SFC}}(i)|^p,$$

and since $p \geq 1$, $x \geq 0$, and $y \geq 0$, we have $|x+y|^{2p} \leq 2^{2p}(x^p + y^p)$, so parallel to our earlier bound but without restricting the X's to post-i variables, we have

$$\begin{aligned}E|d_i|^{2p} &\leq 2^{2p} c_\psi^{2p} EG^p(X_i; X_1, X_2, \ldots, \hat{X}_i, \ldots, X_n) \\ &\quad + 2^{2p} c_\psi^{2p} EG^p(\tilde{X}_i; X_1, X_2, \ldots, \hat{X}_i, \ldots, X_n) \\ &= 2^{2p+1} c_\psi^{2p} EG^p(X_i; X_1, X_2, \ldots, \hat{X}_i, \ldots, X_n).\end{aligned}$$

The last expectation again brings us down to a straightforward calculation about gaps around a point in a random sample of size $n-1$ from the unit interval. After an easy computation, we find

$$EG^p(X_i; X_1, X_2, \ldots, \hat{X}_i, \ldots, X_n) \leq (cp/n)^p,$$

which brings us to the desired bound on the martingale differences,

$$||d_i||_{2p} \leq c(p/n)^{1/2}.$$

Thus the second inequality in the hypothesis of the Lemma 2.7.1 is also proved. By applying Lemma 2.7.2, we conclude the proof of Theorem 2.7.2.

Application to almost sure convergence.

From the last two theorems together with the traditional Borel–Cantelli argument, we find the desired strong law:

$$(2.49) \qquad \lim_{n\to\infty} \frac{L_n^{\text{SFC}}}{\sqrt{n\varphi(\log_p n)}} = 1 \quad \text{a.s.}$$

This limit theorem is less precise than Theorem 2.7.2, but without a rate result for the expectations EL_n^{SFC}, the ratio (2.49) has to go without a rate result. In all events, the most engaging feature of (2.49) is its engaging contrast to the Beardwood–Halton–Hammersley theorem.

2.8. Additional notes.

1. Versions of the Beardwood–Halton–Hammersley theorem for random variables with unbounded support were also developed independently by F. Avram (1991) and A. Yao (1991), who both establish the sufficiency of the condition

$$\int ||x||^\gamma f(x)\,dx < \infty, \quad \text{where } \gamma > (d-1)/d.$$

F. Avram (1991) and A. Yao (1991) also provide results on densities that satisfy additional conditions such as spherical symmetry or a specified rate of decay at infinity. Also, under circumstances where Rhee's sufficient condition is not met, Stadje (1994) has investigated the behavior of the limit superior and limit inferior where one has some modest regularity conditions on the density f. Stadje also uses Laplace's method to provide an interesting refinement of Lemma 2.1.1:

$$\lim_{n\to\infty} n^{1/d} E\left(\min_{1\leq i\leq n} |X_i - x|\right) = f(x)^{-1/d} d^{-1} \pi^{-1/2} \Gamma(1/d) \Gamma((d+2)/d)^{1/d}$$

provided that $f(x) > 0$ and certain regularity conditions apply.

2. The space-filling curve heuristic has a long history. Adler (1986) notes that the heuristic dates back to at least an unpublished work of S. Kakutani in 1966. Still, the idea has become much better known and better understood since the work of Platzman and Bartholdi (1989), and the particular distinction of this later work is that it makes explicit that the required preimage calculations can be done effectively and quickly. Other independent developments of the space-filling curve heuristic can be found in communication theory by Bailey (1969), in matching theory by Imai (1986), and in real analysis by Garsia (1976), Kahane (1976), and Milne (1980). The article of Milne (1980) provides a complete

treatment of the measure-preserving and Lipschitz properties for a class of d-dimensional curves. Milne (1980) also gives an explicit value for the Lipschitz constant of the d-dimensional curves as $3(d+3)^{\frac{1}{2}}$.

3. Few (1962) proved that the constant in (2.13) can be taken to be $(4/3)^{\frac{1}{4}} + \epsilon$ for all $n \geq N(\epsilon)$. Moreover, Few proved that if we take $s_i = s_i(x_1, x_2, \ldots, x_n) = \min\{|x_i - x_j| : j \neq i, 1 \leq j \leq n\}$ and set $x = (x_1, x_2, \ldots, x_n)$, then for $U_n(x) = s_1 + s_2 + \cdots + s_n$ we have $\max_x U_n(x) \sim n^{\frac{1}{2}}(4/3)^{\frac{1}{4}}$ as $n \to \infty$. The best constant in Few's bound is still not known. The articles by Karloff (1989) and Goddyn (1990) give the deepest results to date, and these articles also show just how delicate the bounding problems can be. The article by Snyder and Steele (1995a) studies the sums of the powers of the edge lengths of the TSP, and Snyder and Steele (1995b) explores the extent to which worst-case point sets must echo the behavior of random point sets. Yukich (1995) has recently investigated the worst-case behavior of a large class of geometric problems by means of the method of "rooted duals," which we will discuss in a later chapter.

4. The space-filling curves that we have used here naturally have purely combinatorial, purely discrete counterparts. Alpern (1977) has investigated the labeling of the 2^{kd} squares of the kth-order dyadic decomposition of the unit d-cube and has exhibited a cyclical ordering of the squares such that if i and j are any pair of labels with $|i - j| \leq 2^{d-1} - 2^{d-2} + 2^{d-3}$, then cube i and cube j have a vertex in common. This discrete property is an analogue of the Lipschitz $1/d$ condition that is of importance to us in our applications of the space-filling curves. Garsia (1979) also has useful information on the discrete variants of the space-filling curve problems.

5. Karp (1977) provides a nice algorithmically designed lemma that we would have liked to have used in the proof of inequality (2.32), except that one first needs to introduce a little more terminology from graph theory. We recall that a multigraph is just a graph where one allows multiple edges and loops. Also, a loop at a vertex is understood to contribute two to the degree of that vertex. Further, a multigraph G is weighted if for each edge (x, y) we have a real number $d(x, y)$, and G satisfies the triangle inequality if for any $(x, y), (y, z)$, and (x, z) we have $d(x, z) \leq d(x, y) + d(y, z)$. If H is a subgraph of G, the weight of H is equal to the sum of the weights of the edges of H, and we write $w(H)$ for this sum. Finally, we call a graph G a *spanning walk* for V if G is connected, contains all of the points of V as vertices, and has even degree at each vertex.

LEMMA 2.8.1. *Suppose that G is a spanning walk on the set of vertices V, and suppose the weights on the edges of the complete graph K on V satisfy the triangle inequality. Then there is a tour H of V such that $w(H) \leq w(G)$.*

Proof. The proof uses two operations, each of which takes a spanning walk to a spanning walk without increasing the weight. Each operation also decreases the number of edges in the graph by one, so only finitely many operations are required. Finally, if the spanning walk is not a tour, then we will check that one of the two operations can always be applied. The operations are easy to define:

- LOOP(v). This operation removes a loop at the vertex v.
- PASS(u, v, w). This operation removes edges the edges (u, v) and (v, w) and inserts the edge (u, w).

The operations clearly do not increase the weight and do decrease the number of edges in the graph. They also preserve the spanning and even degree conditions, so we always have a spanning walk. Finally, if the number of edges in G is not exactly equal to card V, then there must be a vertex that has degree at least four. Such a vertex may have a loop, in which case we can use the LOOP operation. If the vertex does not have a loop, then PASS can be applied to the vertex and any two neighbors while still preserving the connectedness of the graph.

CHAPTER 3

More General Methods

This chapter develops a more general view of the class of problems that are typified by the traveling-salesman problem. The resulting theory of subadditive Euclidean functionals turns out to offer a rewarding approach to many concrete problems and also gets closer to the essential features that make possible theorems like that of Beardwood, Halton, and Hammersley. The chapter also reviews recent progress on rates of convergence that have been made possible by the consideration of two-sided bounds.

3.1. Subadditive Euclidean functionals.

We begin by detailing some general properties of a function L from the set of finite subsets of $\mathbb{R}^d$ to the nonnegative real numbers $\mathbb{R}^+$. The intention of these properties is to echo the most basic features of the TSP tour-length function. We first impose a natural normalization,

(3.1) $$L(\emptyset) = 0,$$

and then we consider only the simplest geometric properties of homogeneity and translation invariance:

(3.2) $\quad L(\alpha x_1, \alpha x_2, \ldots, \alpha x_n) = \alpha L(x_1, x_2, \ldots, x_n) \quad$ for all $\alpha > 0$

and

(3.3) $\quad L(x_1 + y, x_2 + y, \ldots, x_n + y) = L(x_1, x_2, \ldots, x_n) \quad$ for all $y \in \mathbb{R}^d$.

There is also a technical property that we require of L. For each n, if we view L as a function from $\mathbb{R}^{nd}$ to $\mathbb{R}$, then we assume that L is Borel measurable. This property is immediate in all of our applications, and, in fact, L usually yields a continuous function on $\mathbb{R}^{nd}$. Functions on the finite subsets of $\mathbb{R}^d$ that are measurable in the sense just described and that satisfy (3.1)–(3.3) are called *Euclidean functionals*.

As they sit, properties (3.1)–(3.3) are still rather toothless, but there is one further property of the TSP that can be joined with the preceding to form a forceful combination. This property already featured directly in the analysis of the expected tour lengths, but since the property is crucial in the abstract setting, it deserves a special christening.

Geometric subadditivity hypothesis. There exists a constant C_0 such that for all integers $m \geq 1$, $n \geq 1$, and $\{x_1, x_2, \ldots, x_n\} \subset [0,1]^d$, we have

$$(3.4) \quad L(\{x_1, x_2, \ldots, x_n\}) \leq \sum_{i=1}^{m^d} L(\{x_1, x_2, \ldots, x_n\} \cap Q_i) + C_0 m^{d-1},$$

where $\{Q_i\}$, $1 \leq i \leq m^d$, is the partition of $[0,1]^d$ into cubes of edge length $1/m$.

On many occasions, geometric subadditivity ((3.4)) can be verified at once for all $m \geq 1$, but in some instances we build our way from the case of $m = 2$ to the general case. For the moment, we just recall that in our analysis of the TSP, we found (3.4) directly for all m at one time, and this situation may be typical. Still, we will see later that there is a benefit to having a few tools around to help prove geometric subadditivity.

In addition to properties (3.1)–(3.4), there is one further property of the TSP functional that proved useful in our earlier analysis. The TSP functional is *monotone* in the sense that for all n and $\{x_i\}$, we have

$$(3.5) \quad L(x_1, x_2, \ldots, x_n) \leq L(x_1, x_2, \ldots, x_n, x_{n+1}).$$

This last property is evident for the TSP, but as we will see shortly, the property is not present in a number of closely related problems that are of considerable importance in the theory of combinatorial optimization. We will revisit this issue of monotonicity in a subsequent section, but for the moment we will exploit monotonicity as best we can.

Euclidean functionals that satisfy (3.4) will be called *subadditive Euclidean functionals*, and the analysis of such functionals is at the heart of this chapter. If (3.5) also holds, we say that L is a *monotone subadditive Euclidean functional*, and this is one particularly simple class of processes that seems to go a long way in capturing the features of the TSP that provide for an effective asymptotic analysis. The main aim of this first section is to show that properties (3.1)–(3.5) are sufficient to determine the asymptotic behavior of $L(X_1, X_2, \ldots, X_n)$, where the $\{X_i\}$'s are independent and uniformly distributed on $[0,1]^d$.

THEOREM 3.1.1 (basic theorem of subadditive Euclidean functionals). *Suppose L is a monotone subadditive Euclidean functional. If the random variables $\{X_i\}$ are independent with the uniform distribution on $[0,1]^d$, then as $n \to \infty$ we have with probability one that*

$$L(X_1, X_2, \ldots, X_n)/n^{(d-1)/d} \to \beta_L,$$

where $\beta_L \geq 0$ is a constant.

Proof. We first check that our assumptions guarantee that the worst-case bound $\|L(X_1, X_2, \ldots, X_n)\|_\infty$ does not grow too rapidly. With even a modest bound on this L^∞ norm, we will be at liberty to use Poissonization and to consider means and variances at will.

The argument for our bound is based on induction on the cardinality of the finite set $F = \{x_1, x_2, \ldots, x_n\} \subset [0,1]^d$. To set up the induction, we first take

$A = C_0 2^{d-1}$, where C_0 is the constant of (3.4), and take $B = A + L(\{x\})$, where we note that by translation invariance ((3.3)), the value of $L(\{x\})$ does not depend on the value of $x \in [0,1]^d$. By these choices, we certainly have that for $F \subset [0,1]^d$ with card $F = 1$,

(3.6) $$L(F) \leq B \operatorname{card}(F) - A.$$

We take (3.6) as our induction hypothesis, and we assume that (3.6) holds for all $F \subset [0,1]^d$ with $1 \leq \operatorname{card} F < n$.

We now consider the partition of $[0,1]^d$ into 2^d equal subcubes Q_i with edge length $\frac{1}{2}$, and we let $F \subset [0,1]^d$ be any set with card $F = n$. By the translation property (3.3), we can assume without loss of generality that F is not contained in any one of the Q_i's, so if we let $I = \{i : F \cap Q_i \neq \emptyset\}$, we can assume that card $I \geq 2$. By geometric subadditivity (3.4) and the induction hypothesis, we therefore find that

(3.7) $$L(F) \leq \sum_{i \in I} L(F \cap Q_i) + C_0 2^{d-1} \leq Bn - 2A + C_0 2^{d-1} = Bn - A,$$

so (3.7) completes the proof of the induction step.

Now we are in a position to take advantage of Poissonization in a way that parallels our analysis of the TSP. We let Π denote the Poisson process in $\mathbb{R}^d$ with unit intensity, and we set $Z(t) = L(\Pi[0,t]^d)$. We note that by our induction argument we have $Z(t) = O(\operatorname{card}(\Pi[0,t]^d))$, so $Z(t)$ has moments of all orders.

We first work toward showing that $EZ(t)/t^d$ converges, and for a while, the argument closely parallels one from the previous chapter. By (3.4) applied to $\Pi[0,t]^d$, we again get $EZ(t) \leq m^d EZ(t/m) + C_0 t m^{d-1}$, which on replacing t by mt and dividing by $m^d t^d$ gives us the key relation:

(3.8) $$EZ(mt)/(m^d t^d) \leq EZ(t)/t^d + C_0 t^{1-d}.$$

We can then define γ by

(3.9) $$0 \leq \gamma \equiv \liminf_{m \to \infty} EZ(m)/m^d \leq EZ(1) + C_0 < \infty.$$

Now for the last part of the argument that parallels our analysis of the TSP, we note that given any $\epsilon > 0$, we can choose a t_0 such that

$$EZ(t_0)/t_0^d + C_0 t_0^{1-d} \leq \gamma + \epsilon.$$

The argument now changes at least a little as we start to rely more on the monotonicity property (3.5) of L, though the significant changes emerge only when we look at the second moment of $Z(t)$ and when we need to back out of the Poissonization.

We first note that the monotonicity (3.5) of L gives us the pathwise monotonicity of $Z(t)$, so for $mt_0 \leq u < (m+1)t_0$ we certainly have the expectation bounds

$$EZ(u)/u^d \leq EZ((m+1)t_0)/(mt_0)^d \leq (\gamma + \epsilon)(m+1)^d/m^d.$$

Hence we find by taking the limit supremum and using the arbitrariness of m that
$$\limsup_{u\to\infty} EZ(u)/u^d \le \liminf_{t\to\infty} EZ(t)/t^d + \epsilon,$$
so by the arbitrariness of $\epsilon > 0$, we conclude that

(3.10) $$EZ(t)/t^d \to \gamma \quad \text{as } t \to \infty.$$

The next step is to work toward an understanding of the second moment of $Z(t)$, and finally the argument sets a new course. We begin by applying the geometric subadditivity condition (3.4) with $m = 2$ to the Poisson sample $\Pi[0, 2t]^d$. On writing $Z_i(t)$ for $L(\Pi(Q_i))$ and noting the change of scale, we find

(3.11) $$Z(2t) \le \sum_{i=1}^{2^d} Z_i(t) + C_0 2^{d-1} t.$$

To get a somewhat simpler inequality, we set $\hat{Z}(t) = Z(t) + 2C_0 t$ and $\hat{Z}_i(t) = Z_i(t) + 2C_0 t$ so that inequality (3.11) implies

(3.12) $$\hat{Z}(2t) \le \sum_{i=1}^{2^d} \hat{Z}_i(t).$$

If we write $\phi(t) = E\hat{Z}(t) = E\hat{Z}_i(t)$ and $\psi(t) = (E\hat{Z}(t)^2)^{\frac{1}{2}} = (E\hat{Z}_i(t)^2)^{\frac{1}{2}}$, we find by squaring (3.12) and taking expectations that
$$\psi^2(2t) \le 2^d \psi^2(t) + 2^d(2^d - 1)\phi^2(t).$$

Introducing $V(t) = \text{Var}\, Z(t) = \psi^2(t) - \phi^2(t)$, we are led to
$$V(2t) = \psi^2(2t) - \phi^2(2t) \le 2^d V(t) + 2^{2d}\phi^2(t) - \phi^2(2t),$$
and upon dividing by $(2t)^{2d}$, we find our fundamental recursion:

(3.13) $$\frac{V(2t)}{(2t)^{2d}} - \frac{V(t)}{2^d t^{2d}} \le \frac{\phi^2(t)}{t^{2d}} - \frac{\phi^2(2t)}{(2t)^{2d}}.$$

By applying this bound for $t, 2t, \ldots, 2^{M-1}t$ and summing, we find
$$\sum_{k=1}^{M} \frac{V(2^k t)}{(2^k t)^{2d}} - 2^{-d} \sum_{k=0}^{M-1} \frac{V(2^k t)}{(2^k t)^{2d}} \le \frac{\phi^2(t)}{t^{2d}},$$
so finally,
$$(1 - 2^{-d}) \sum_{k=1}^{M} \frac{V(2^k t)}{(2^k t)^{2d}} \le \frac{\phi^2(t)}{t^{2d}} + \frac{V(t)}{t^{2d}}.$$

Since this bound holds for all M, we can let $M \to \infty$ and arrive at the main fact we need concerning $V(t)$:

(3.14) $$\sum_{k=1}^{\infty} V(2^k t)/(2^k t)^{2d} < \infty.$$

In order to use (3.14) to complete the proof, we again call on the Borel–Cantelli lemma, but this time we need to add an interpolation argument that makes essential use of monotonicity; after all, the sum (3.14) is over a very sparse subsequence.

The argument first requires that we bring our original variables $\{X_1, X_2, \ldots\}$ back into view. To do this, we let $N(t)$ be a regular (one-dimensional) Poisson counting process with rate one that is independent of $\{X_i : 1 \leq i < \infty\}$. The point of this change is the simple observation that $Z(t) = L(\Pi[0,t]^d)$ and $tL(X_1, X_2, \ldots, X_{N(t^d)})$ have the same distribution for each t, but we have made progress since the $\{X_i\}$'s of our theorem are present in the second expression. Since we have already found that $EZ(t) \sim \beta t^d$ as $t \to \infty$, we find from (3.14) and Chebyshev's inequality that for any $\epsilon > 0$ we have

$$\sum_{k=0}^{\infty} P\{|t2^k L(X_1, X_2, \ldots, X_{N(t^d 2^{kd})})/(t2^k)^d - \beta| > \epsilon\} < \infty.$$

By the Borel–Cantelli lemma, we see that for each $t > 0$,

(3.15) $$\lim_{k \to \infty} L(X_1, X_2, \ldots, X_{N(t^d 2^{kd})})/(t2^k)^{d-1} = \beta \quad \text{a.s.}$$

Now for the interpolation argument. For any fixed integer $p > 0$ and any real number $s \geq 2^p$, we can find integers t and k such that $2^p \leq t < 2^{p+1}$ and $2^k t \leq s \leq 2^k(t+1)$. By the monotonicity of L, we then have

$$L(X_1, X_2, \ldots, X_{N(t^d 2^{kd})}) \leq L(X_1, X_2, \ldots, X_{N(s^d)}) \leq L(X_1, X_2, \ldots, X_{N((t+1)^d 2^{kd})}).$$

Since p is fixed, the set of integers $\{t : 2^p \leq t < 2^{p+1}\}$ is finite; the last pair of inequalities and the subsequence limit (3.15) imply that with probability one we have for real $s \to \infty$ that

$$\limsup_{s \to \infty} L(X_1, X_2, \ldots, X_{N(s^d)})/s^{d-1} \leq \beta(1 + 2^{-p})^{d-1}$$

and

$$\liminf_{s \to \infty} L(X_1, X_2, \ldots, X_{N(s^d)})/s^{d-1} \geq \beta(1 + 2^{-p})^{1-d}.$$

By the arbitrariness of the integer p, we see as $s \to \infty$ that

(3.16) $$\lim_{s \to \infty} L(X_1, X_2, \ldots, X_{N(s^d)})/s^{d-1} = \beta \quad \text{a.s.}$$

The last step we take is to back out of the Poisson indexing. To do so, we let $\tau(n) = \min\{t : N(t^d) = n\}$ and note that by the definition of $\tau(n)$ we have the identity

(3.17) $$L(X_1, X_2, \ldots, X_n)/n^{(d-1)/d}$$
$$= \{L(X_1, X_2, \ldots, X_{N(\tau(n)^d)})/\tau(n)^{d-1}\}\{\tau(n)^{d-1}/n^{(d-1)/d}\}.$$

Finally, the first factor above goes to β with probability one by (3.16), and a well-known property of the Poisson process is that $\tau(n)/n^{1/d} \to 1$ with probability one; so the second factor also converges to one almost surely, and we see that the proof of the theorem is complete.

One aspect of the proof may bother individuals with a measure-theoretically suspicious nature. With the introduction of the process $\{N(t)\}$, we enlarge the probability space and therefore change the meaning of "almost sure." There is really no problem in this case since N only drives n to infinity; and, as the first term of (3.17) shows, N does not enter into our final assertion. Still, some meditation on this point would no doubt be good for the soul, and perhaps a word or two about Fubini's theorem might help as well.

Finally before leaving Theorem 3.1.1, we should note where the monotonicity hypothesis was important. We did use it in the derivation of (3.10), which gives the asymptotics of the Poissonized mean, though the use of the monotonicity there is rather minor and with only small changes the use of monotonicity could have been avoided at that point. The key role for monotonicity was rather in the passage from (3.14) to the convergence in (3.16).

Shortly, we will develop flexible alternatives to the monotonicity hypothesis in order to provide results that are readily applied. There is a certain charm to the fact that Theorem 3.1.1 is purely one sided in the sense that (3.5) and (3.4) provide constraints on $L(x_1, x_2, \ldots, x_n)$ only from above. This fact is in keeping with the structure of Kingman's subadditive ergodic theorem as well as with other results of subadditive theory. Still, there is both flexibility and forcefulness in two-sided constraints, and we will see in several instances that two-sided constraints can yield results that one-sided constraints seem powerless to provide.

3.2. Examples: Good, bad, and forthcoming.

Theorem 3.1.1 was designed with the TSP in mind, so it comes as no surprise that it contains the uniform case of the Beardwood–Halton–Hammersley theorem. More to the point, Theorem 3.1.1 also includes many other problems that are related to the TSP, like the TSP in which the cost of an edge $e = (x, y)$ in the tour is given by the L^p norm on $\mathbb{R}^d$, $p \geq 1$,

$$c(e) = c(x, y) = \left(\sum (x_i - y_i)^p \right)^{1/p},$$

or the L^∞ norm on $\mathbb{R}^d$, where $c(e) = c(x, y) = \max |x_i - y_i|$.

A more interesting example that is still clearly within the domain of Theorem 3.1.1 is the Steiner tree problem. In this problem, we are given a set of points $\{x_1, x_2, \ldots, x_n\} \subset [0,1]^d$, and we are asked to determine a tree that *contains* the set and which has the minimum Euclidean length of any such tree. This problem should be contrasted with the minimum-spanning-tree (MST) problem, which asks instead for the spanning tree of $\{x_1, x_2, \ldots, x_n\}$ that has minimum Euclidean length. The lengths associated with both of these problems are subadditive Euclidean functionals, but the Steiner tree functional is monotone while the MST functional is not monotone, as one can see by consideration of the set consisting of the four corners of the unit square and the set of five points that adds the center point of the square. The MST of the four set has length 3 while the MST of the five set has length $2\sqrt{2}$.

After some consideration of the MST and other examples, the limitations of the monotonicity hypothesis of Theorem 3.1.1 start to become evident, and one starts to look for ways to soften that assumption. Work in this direction was begun in Steele (1981a), but now there are much better approaches. The first of these that we will explore is due to Rhee (1993a).

3.3. A general L^∞ bound.

In all of the examples of subadditive Euclidean functionals that have been explored by direct methods, investigators have always found that the expected growth rate of $O(n^{(d-1)/d})$ is complemented by a similar bound on the L^∞ norm. This phenomenon turns out to have been in the cards, as the following result confirms.

LEMMA 3.3.1 (Rhee (1993a)). *For any Euclidean functional with geometric subadditivity (3.4) for $m = 2$, there is a constant C_1 such that for any finite $F \subset [0,1]^d$ we have*

$$(3.18) \qquad L(F) \leq C_1 (\operatorname{card} F)^{(d-1)/d}.$$

Proof. The proof proceeds by induction, and as one often finds in such arguments, we need a slightly stronger induction hypothesis than the one that appears most naturally in the theorem. We first note by translation invariance that $L(\{x\}) = a$ does not depend on x. Next, to help frame the induction hypothesis, we set $a_1 = a + d2^{d-1}a_2$, where we take $a_2 = C_0/(2^{d-1} - 1)$ and C_0 is the constant that appears in (3.4), the geometric subadditivity inequality for L. Finally, as our induction hypothesis, we will assume that for all $F \subset [0,1]^d$ with cardinality less than n, we have that

$$(3.19) \qquad L(F) \leq a_1 (\operatorname{card} F)^{(d-1)/d} - a.$$

By the definition of a and a_2, the hypothesis is true for F with cardinality one, so our induction starts out on the right foot.

Now consider the usual partition of $[0,1]^d$ into 2^d equal cubes Q_i. By translation invariance and homogeneity, we can assume that F is not contained in any one of the Q_i's. If we write $n_i = \operatorname{card}(F \cap Q_i)$, then for all i we have $n_i < n$, and our induction hypothesis tells us that for all i with $n > n_i > 0$, we have

$$L(F \cap Q_i) \leq \frac{1}{2}\left(a_1 n_i^{(d-1)/d} - a_2\right).$$

Since $L(\emptyset) = 0$, we have $L(F \cap Q_i) = 0$ when $n_i = 0$, so we can set $I = \{i : n_i > 0\}$ and note by geometric subadditivity that

$$L(F) \leq C_0 + \frac{1}{2} \sum_{i \in I} \left(a_1 n_i^{(d-1)/d} - a_2\right).$$

If we set $k = \operatorname{card} I$, then Jensen's inequality gives us

$$L(F) \leq C_0 + \frac{1}{2} a_1 k^{1/d} n^{(d-1)/d} - \frac{1}{2} a_2 k.$$

We need to show that the right-hand side is bounded by $a_1 n^{(d-1)/d} - a_2$; or, after substituting $C_0 = a_2(2^{d-1} - 1)$ and using $n \geq 1$ and $a_1 \geq d2^{d-1} a_2$, we just need to show that
$$2^d - k \leq d2^d(1 - k^{1/d}/2).$$
If we choose x so that $k = x2^d$, a proof of the last inequality reduces to showing that $1 - x \leq d(1 - x^{1/d})$ for $0 \leq x \leq 1$. This elementary fact follows from the convexity of the curve $y = d(1 - x^{1/d})$ and the tangency of this curve to the line $y = 1 - x$ at the point $x = 1$.

3.4. Simple subadditivity and geometric subadditivity.

The geometric-subadditivity property seems to be at the heart of the theory of subadditive Euclidean functionals, but there is a more elementary relationship that one finds in virtually all of the problems where one has geometric subadditivity (3.4). This is the property of *simple subadditivity*, by which we mean that for disjoint finite sets F_1 and F_2 of $[0,1]^d$, we have

(3.20) $$L(F_1 \cup F_2) \leq L(F_1) + L(F_2) + C_2.$$

By $2^d - 1$ applications of (3.20), we easily find that (3.20) implies the $m = 2$ case of (3.4) with $C_0 = C_2(2^d - 1)/2^{d-1}$. This modest fact immediately provides a useful corollary to the Rhee's lemma (Lemma 3.3.1). In fact, we get the same conclusion with the modest benefit that only simple subadditivity needs to be introduced.

LEMMA 3.4.1. *For any Euclidean functional with simple subadditivity* (3.20), *there is a constant* C_1 *such that for any finite* $F \subset [0,1]^d$ *we have*

(3.21) $$L(F) \leq C_1 (\text{card } F)^{(d-1)/d}.$$

By the same argument applied for the case $m = 2$ to prove Lemma 3.4.1, one can prove geometric subadditivity (3.4) for all $m \leq M$ for any *fixed* M. Since the constant one obtains in this way depends on M, one cannot use this naïve argument to extract the full geometric subadditivity (3.4) from (3.20). Nevertheless, we will shortly provide an argument that suggests how one can often prove (3.4) with little more than (3.20), though apparently there remains a need for some additional facts for the functional L that is at hand. One interesting consequence of Lemma 3.4.1 is the implication that a monotone Euclidean functional with simple subadditivity must also satisfy a strong smoothness condition.

LEMMA 3.4.2. *For any Euclidean functional L that is monotone* ((3.5)) *and simply subadditive, there is a constant C_3 such that for all finite subsets F and G of $[0,1]^d$, we have*
$$|L(F \cup G) - L(F)| \leq C_3 (\text{card } G)^{(d-1)/d}.$$

Proof. By monotonicity and simple subadditivity, we have
$$L(F) \leq L(F \cup G) \leq L(F) + L(G) + C_2,$$
so when we bound $L(G)$ using Lemma 3.4.1, we find

(3.22) $$L(F) \leq L(F \cup G) \leq L(F) + (C_1 + C_2)(\text{card } G)^{(d-1)/d}.$$

This lemma would be a delight, except one of the practical realizations to emerge after considerable investigation into Euclidean functionals is that the continuity condition (3.22) is actually more convenient in applications than the apparently simpler monotonicity. The next section pursues this theme by illustrating several ways that the smoothness condition (3.22) works nicely together with Azuma's inequality.

3.5. A concentration inequality.

The Hamming distance H on a product space Ω^n measures the distance between x and y by the number of coordinates in which x and y disagree, so if $H(x,y)$ is the Hamming distance between x and y, we have

$$d_H(x,y) = \text{card}\{i : x_i \neq y_i\}.$$

As an easy consequence of Azuma's inequality, we can prove that in many cases all but a small part of Ω^n is within a short Hamming distance of $A \subset \Omega^n$ provided that the set A is not too small. This is an instance of what people nowadays refer to as an isoperimetric inequality, the theory of which will be of great concern to us in the last chapter.

LEMMA 3.5.1 (isoperimetric inequality for Hamming distance). *Let* $\Omega = [0,1]^d$ *and let* μ *denote the uniform measure on the n-fold product space* Ω^n. *For a* $A \subset \Omega^n$, *we define* $\phi_A(y) = \min\{k : \exists x \in A \text{ with } H(x,y) \leq k\}$. *If* $\mu(A) \geq \frac{1}{2}$, *then*

$$(3.23) \qquad \mu(\{y : \phi_A(y) \geq t\}) \leq 4\exp(-t^2/8n).$$

Proof. Since changing one of the coordinates of $y = (y_1, y_2, \ldots, y_i, \ldots, y_n)$ can only change $H(x,y)$ by at most one for any x, we clearly have $|\phi_A(y_1, y_2, \ldots, y_i, \ldots, y_n) - \phi_A(y_1, y_2, \ldots, y_i', \ldots, y_n)| \leq 1$. Thus, Azuma's inequality applied to the martingale-difference representation of $\phi_A(y)$ gives us

$$(3.24) \qquad \mu(|\phi_A(y) - \alpha| \geq u) \leq 2\exp(-u^2/2n),$$

where $\alpha = \int \phi_A(y) d\mu$. If we let $u = \alpha$ in (3.24), then since $\phi_A(y) = 0$ for $y \in A$, we see that the left-hand side of (3.24) is at least as large as $\frac{1}{2}$, so $\frac{1}{2} \leq 2\exp(-\alpha^2/2)$, or $\alpha \leq (2n \log 4)^{\frac{1}{2}}$. Applying this bound on α in (3.24) then gives us for all $u \geq 0$ that

$$\mu\left(\phi_A(y) \geq u + \sqrt{(2n \log 4)}\right) \leq 2\exp(-u^2/2n),$$

so for $t \geq (2n \log 4)^{\frac{1}{2}}$ we have

$$(3.25) \qquad \mu(\phi_A(y) \geq t) \leq 2\exp\left(-\left(t - \sqrt{(2n \log 4)}\right)^2/2n\right).$$

Now since $(t - a)^2 \geq t^2/4$ for $t \geq 2a$, if we take $t \geq 2(2n \log 4)^{\frac{1}{2}}$ in (3.25), we find

$$\mu(\phi_A(y) \geq t) \leq 2\exp(-t^2/8).$$

Finally, since for $0 \leq t \leq 2(2n\log 4)^{\frac{1}{2}}$ we have $2\exp(-t^2/8) \geq \frac{1}{2}$, if we multiply the right-hand side of the last inequality by 2, we have an inequality that is valid for all $0 \leq t < \infty$, and the proof of the lemma is complete.

The meaning of the lemma is illuminated by consideration of the sets

$$A_t = \{x \in \Omega^n : H(x,y) \leq t \text{ for some } y \in A\}.$$

The sets A_t are fattened versions of A that also include all of the points of Ω^n that are within a Hamming distance t of A. The lemma then says that if $\mu(A) \geq \frac{1}{2}$ and $t \gg \sqrt{n}$, then the measure of A_t is almost 1 in the sense that

$$\mu(A_t) \geq 1 - 4\exp(-t^2/8n).$$

In order to apply the preceding bounds in the theory of Euclidean functionals, we focus on functionals that satisfy a natural smoothness condition that we find to be satisfied by a large number of subadditive Euclidean functionals. In fact, we have already seen that one half of the smoothness inequality is automatically satisfied for any Euclidean functional with either the geometric subadditivity property or simple subadditivity.

THEOREM 3.5.1 (Rhee (1993a)). *If L is a measurable function on the finite subsets of $[0,1]^d$ and if L satisfies the smoothness condition*

(3.26) $$|L(F \cup G) - L(F)| \leq C_3 (\text{card } G)^{(d-1)/d}$$

for all finite subsets F and G of $[0,1]^d$, then there is a universal $C = C(L,d)$ such that for independent random variables $\{X_i\}$ with the uniform distribution on $[0,1]^d$, one has

$$P\left(|L(\{X_1, X_2, \ldots, X_n\}) - EL(\{X_1, X_2, \ldots, X_n\})| \geq t\right)$$

(3.27) $$\leq C \exp\left(-\frac{1}{Cn}\left(\frac{t}{C_3}\right)^{2d/(d-1)}\right).$$

Proof. We let $Z = L(\{X_1, X_2, \ldots, X_n\})$, and we set M equal to the median of Z. If $A = \{x = (x_1, x_2, \ldots, x_n) : L(x) \leq M\}$, then we note by the definition of ϕ_A that for each $y = (y_1, y_2, \ldots, y_n)$ there is an $x \in A$ such that $H(x,y) \leq \phi_A(y)$. Given y, we then set $F = \{x_i : x_i = y_i\}$ and $G = \{x_i : x_i \neq y_i\}$, and we note that from

$$L(y) = \{L(y) - L(F)\} + \{L(F) - L(x)\} + L(x),$$

we find

$$L(y) \leq 2C_3 \phi_A(y)^{(d-1)/d} + M$$

since $L(x) \leq M$ and by (3.26) we know both $|L(y) - L(F)|$ and $|L(F) - L(x)|$ are bounded by $C_3(\text{card } G)^{(d-1)/d}$.

We therefore find from Lemma 3.5.1 that

$$P(Z \geq M + t) \leq P\left(\phi_A(y)^{(d-1)/d} \geq (t/C_3)^{(d-1)/d}\right)$$

(3.28) $$\leq 4\exp\left(-\frac{1}{8n}\frac{t^{2d/(d-1)}}{(2C_3)^{2d/(d-1)}}\right).$$

Since the same argument can be applied to $P(Z \leq M + t)$, we find that $P(|Z - M| \geq t)$ is bounded by twice the last term of (3.28). By integration of this tail bound, we further find $E|M - Z| \leq Kn^{(d-2)/2d}$, and by taking the absolute values outside of the expectation, we get $|M - EZ| \leq Kn^{(d-2)/2d}$. This estimate of EZ can be applied in (3.28) to give us

$$(3.29) \quad P\left(|Z - EZ| \geq t + Kn^{(d-2)/2d}\right) \leq 8\exp\left(-\frac{1}{8n}\frac{t^{2d/(d-1)}}{(2C_3)^{2d/(d-1)}}\right).$$

Now since $(t - a)^{2d/(d-1)} \geq (t/2)^{2d/(d-1)}$ for $t \geq 2a$, we see for $t \geq 2Kn^{(d-2)/2d}$ that inequality (3.29) gives us

$$(3.30) \quad P(|Z - EZ| \geq t) \leq 8\exp\left(-\frac{1}{8n}\frac{t^{2d/(d-1)}}{(2C_3)^{2d/(d-1)}}\left(\frac{1}{2}\right)^{2d/(d-1)}\right).$$

Finally, for all $0 \leq t \leq 2Kn^{(d-2)/2d}$ and all $n \geq 1$, the right-hand side of (3.30) is bigger than a constant $0 < K' = K'(L, d) < 1$, so on combining the bounds for the two ranges, we see that for all $0 \leq t < \infty$ we have that $P(|Z - EZ| \geq t)$ is no bigger than the right-hand term of (3.30) divided by K'.

The hypotheses of Rhee's concentration inequality are easily verified in many problems of geometric combinatorial optimization. Also, with the typical behavior of the mean that one finds for subadditive Euclidean functionals, $EL_n \sim cn^{(d-1)/d}$, Rhee's concentration inequality leads to a strong law like that of the Beardwood–Halton–Hammersley theorem. In fact, we get even more since without any further work, Rhee's inequality gives us complete convergence. One of the other benefits of Rhee's inequality is that it leads us to an elegant treatment of the minimal-matching problem—the topic of the next section.

3.6. Minimal matching.

The Euclidean minimal-matching problem considers a set of even cardinality $\{x_1, x_2, \ldots, x_{2n}\}$ in $[0,1]^d$ and asks for a way to join these points together to form a set of disjoint pairs so that the sum of the distances between the paired points is minimal. The resulting length $L(\{x_1, x_2, \ldots, x_{2n}\})$ can also be expressed as

$$(3.31) \quad \min_{\sigma}\{\|x_{\sigma(1)} - x_{\sigma(n+1)}\| + \|x_{\sigma(2)} - x_{\sigma(n+2)}\| + \cdots + \|x_{\sigma(n)} - x_{\sigma(2n)}\|\},$$

where the minimum is over all permutations of $[2n] = \{1, 2, \ldots, 2n\}$. Also, to lessen our consideration of special cases, we take the usual convention and define L on sets of odd cardinality by allowing one point to go unmatched.

We first note that the minimal-matching functional satisfies the simple subadditivity property (3.20). To see this, note that if F_1 and F_2 are disjoint finite sets and if one or both of F_1 and F_2 have even cardinality, then by suboptimality we have $L(F_1 \cup F_2) \leq L(F_1) + L(F_2)$. Similarly, if both F_1 and F_2 have odd cardinality, then by joining the unmatched points in the optimal matchings of F_1 and F_2, we have $L(F_1 \cup F_2) \leq L(F_1) + L(F_2) + \sqrt{d}$. We therefore find that we have (3.20) with $C_2 = \sqrt{d}$.

We will next see that the minimal-matching functional is smooth in the sense that we have (3.26); that is, there is a constant C_3 such that for all finite subsets F and G of $[0,1]^d$, we have

(3.32) $$|L(F \cup G) - L(F)| \leq C_3(\operatorname{card} G)^{(d-1)/d}.$$

As our first step toward the proof of this inequality, we note that by simple subadditivity and Lemma 3.3.1, we find

(3.33) $$L(F \cup G) \leq L(F) + L(G) + C_2 \leq L(F) + (C_1 + C_2)(\operatorname{card} G)^{(d-1)/d},$$

which proves one half of the desired smoothness bound.

For the other half, we first consider an optimal matching of $F \cup G$. We let F' be the set of points of F that are matched to points of G, and we let F'' be the complementary set of points in F. Taking $F' = F_1$ and $F'' = F_2$ in the simple subadditivity relation (3.20), we find

$$L(F) = L(F' \cup F'') \leq L(F') + L(F'') + C_2 \leq C_1(\operatorname{card} G)^{(d-1)/d} + L(F'') + C_2,$$

where to bound $L(F')$, we first used Lemma 3.3.1 and then used the fact that since all points of F' are matched with points of G, we have $\operatorname{card} F' \leq \operatorname{card} G$. We next note that the edges of an optimal matching of F'' are also edges of an optimal matching of $F \cup G$ since the restriction of an optimal matching is an optimal matching; so we have $L(F'') \leq L(F \cup G)$, and the preceding inequality can be written as

(3.34) $$L(F) - C_1 - C_2(\operatorname{card} G)^{(d-1)/d} \leq L(F \cup G).$$

Thus by (3.33) and (3.34), the smoothness inequality for the minimal-matching functional has been proved with $C_3 = C_1 + C_2$.

Geometric subadditivity from simple subadditivity.

We will now check that minimal matching satisfies the fundamental geometric subadditivity condition (3.4). That is, we will show that there is a constant C_0 such that for all integers $m \geq 1$, $n \geq 1$, and $\{x_1, x_2, \ldots, x_n\}$, we have

(3.35) $$L(\{x_1, x_2, \ldots, x_n\}) \leq \sum_{i=1}^{m^d} L(\{x_1, x_2, \ldots, x_n\} \cap Q_i) + C_0 m^{d-1},$$

where $\{Q_i\}$, $1 \leq i \leq m^d$, is a partition of $[0,1]^d$ into cubes of edge length $1/m$. We already know by Lemma 3.4.1 that there is a C_1 such that $L(F) \leq C_1(\operatorname{card} F)^{(d-1)/d}$ for all finite $F \subset [0,1]^d$. Now we take the union U of the m^d optimal matchings of the set $F_i = \{x_1, x_2, \ldots, x_n\} \cap Q_i$ and note that the set F of points of $\{x_1, x_2, \ldots, x_n\}$ that are not covered by U has cardinality bounded by m^d. Since U together with a matching of F provides a matching of $\{x_1, x_2, \ldots, x_n\}$, the bound on $L(F)$ guaranteed by Lemma (3.3.1) completes the proof of (3.35) for general m.

This is an argument of notable generality, and even though the argument is not bundled into a tidy proposition, the punchline is clear—whenever one tries to prove a bound like (3.35), Lemma 3.4.1 serves as a trusty tool that shows how cheaply one can patch together subcube solutions to get a solution for the whole cube. In many investigations, considerable art has been expended in the ad hoc derivation of such bounds, and, while the present approach does not guarantee that such art will not be required in the future, it does tell us that many past troubles could have been avoided.

Limit result for minimal matching.

By geometric subadditivity (3.35) and smoothness (3.32), we can repeat our argument of the first section to prove that for Poisson $N = N(\lambda)$ with mean λ, the minimum-matching function satisfies

$$(3.36) \qquad \frac{EM_N}{\lambda^{(d-1)/d}} \to \gamma_{\mathrm{MM}}(d) > 0$$

as $\lambda \to \infty$. Since the smoothness condition (3.32) is more than we need for direct de-Poissonization, we also have the corresponding statement for M_n, the minimal-matching length for an independent uniform sample of size n.

3.7. Two-sided bounds and first consequences.

Naturally, one would like to extract even more from Rhee's concentration inequality. For example, one surely expects the inequality to provide an almost sure rate result for the difference $EM_n - \gamma_{\mathrm{MM}}(d)n^{(d-1)/d}$. The bottleneck is that one first needs a rate result for the mean itself; that is, one has to replace $EM_n \sim -\gamma_{\mathrm{MM}}(d)n^{(d-1)/d}$ with, say, $EM_n = \gamma_{\mathrm{MM}}(d)n^{(d-1)/d} + O(n^{(d-2)/d})$. Pleasantly, this can be done for the minimal-matching problem and for many others. The essential step is to complement the majorization given by geometric subadditivity with a lower bound. As observed in Rhee and Talagrand (1989) and Jaillet (1992), such two-sided bounds can be used to provide rate results in a variety of problems. In fact, two-sided bounds turn out to be important in several aspects of the theory of subadditive Euclidean functionals. For example, we make essential use of two-sided bounds in the extension of the Beardwood–Halton–Hammersley theorem from uniform to general densities. For comparison with the more refined bounds that we will develop to get good rate results, we first recall the pair of inequalities that we use in our argument for the nonuniform Beardwood–Halton–Hammersley theorem. We will see shortly that in $d = 2$, these bounds are good enough to provide rate results that are best possible. For $d > 2$, we will need to develop new tools, but exploration of the case of $d = 2$ will provide us with a useful guide.

LEMMA 3.7.1. *There is a constant $c(d,m) > 0$ depending only on d and m such that for any partition $\{Q_i\}$ of $[0,1]^d$ into m^d subcubes of side $1/m$, we have geometric subadditivity*

$$(3.37) \qquad L^{\mathrm{TSP}}(X_1, X_2, \ldots, X_n)$$

$$(3.38) \qquad \leq \sum_{i=1}^{m^d} L^{\mathrm{TSP}}(\{X_i : X_i \in Q_i\}) + c(d,m),$$

and a complementary inequality

$$\text{(3.39)} \quad \sum_{i=1}^{m^d} L^{\text{TSP}}(\{X_i : X_i \in Q_i\})$$

$$\leq L^{\text{TSP}}(X_1, X_2, \ldots, X_n) + c(d,m)n^{(d-2)/(d-1)}.$$

The case $d = 2$ of this inequality is especially forceful since the term $n^{(d-2)/(d-1)}$ reduces to a constant. If we apply the inequality for $d = 2$ to $\{X_1, X_2, \ldots, X_N\}$, where N is Poisson with mean t, and if we let $\phi(t) = EL_N^{\text{TSP}}$, then we can set $c = c(2,2)$ and note by scaling that for all $t > 0$,

$$\left| \phi(t) - 4\phi(t/4) \cdot \frac{1}{2} \right| \leq c.$$

If we replace t by $4t$ and divide by $\sqrt{4t}$, we find

$$\left| \frac{\phi(4t)}{\sqrt{4t}} - \frac{\phi(t)}{\sqrt{t}} \right| \leq \frac{c}{2\sqrt{t}},$$

which is an inequality that is ripe for iteration if there ever was one. If we replace t by $4^k t$, we find

$$\left| \frac{\phi(4^{k+1}t)}{\sqrt{4^{k+1}t}} - \frac{\phi(4^k t)}{\sqrt{4^k t}} \right| \leq \frac{c}{2^k \sqrt{t}},$$

and we can sum these inequalities for $k = 0, 1, 2, \ldots, m$ to find

$$\left| \frac{\phi(4^m t)}{\sqrt{4^m t}} - \frac{\phi(t)}{\sqrt{t}} \right| \leq \frac{c}{\sqrt{t}}.$$

By the arbitrariness of m, the convergence $\phi(4^m t)/\sqrt{4^m t} \to \beta_{\text{TSP}}(2)$ tells us that we have a nice rate result for the Poissonized TSP in $d = 2$:

$$\text{(3.40)} \quad \left| \frac{\phi(t)}{\sqrt{t}} - \beta_{\text{TSP}}(2) \right| \leq \frac{c}{\sqrt{t}}.$$

For a quick and crude de-Poissonization, we let N denote a Poisson random variable with mean n and note that from our L^∞ bound for the TSP, $|L_N^{\text{TSP}} - L_n^{\text{TSP}}| \leq c(N-n)^{1/2}$, we have

$$|\phi(n) - EL_n^{TSP}| \leq E|L_N^{\text{TSP}} - EL_n^{\text{TSP}}|$$

$$\leq cE\left\{(N-n)^{1/2}\right\} \leq c\left(E(N-n)^2\right)^{1/4} = cn^{1/4}.$$

This inequality tells us that $EL_n^{\text{TSP}} - \phi(n) = O(n^{\frac{1}{4}})$, and our iteration gave us $\phi(n)/\sqrt{n} = \beta_{\text{TSP}}(2) + O(n^{-\frac{1}{2}})$, so we have

$$\text{(3.41)} \quad EL_n^{\text{TSP}}/\sqrt{n} - \beta_{\text{TSP}}(2) = O(n^{-1/4}).$$

This particular de-Poissonization argument was given just to see how it might be improved—a proverbial strawman, built to be knocked down. Clearly, we have not made the best use of the elegant Poisson rate result (3.40), and our troubles come from our crude use of the L^∞ bound, $|L_N^{\text{TSP}} - L_n^{\text{TSP}}| \leq c(N-n)^{\frac{1}{2}}$. This inequality is efficient if N and n differ by a good fraction of n, but for small differences between n and N, the bound is too far off the mark. Fortunately, we have information at our disposal that will let us extract the best from (3.40), and for purposes of later comparison, we frame the required fact as a lemma.

LEMMA 3.7.2 (add-one bound). *For the TSP functional applied to an independent uniformly distributed sample from the unit square $[0,1]^2$, we have a constant $c > 1$ such that for all $n > 1$,*

$$(3.42) \qquad \left| EL_n^{\text{TSP}} - EL_{n+1}^{\text{TSP}} \right| \leq \frac{c}{\sqrt{n}}.$$

The lemma is a special case of our earlier work, so there is no need to give a proof. The more interesting bit is the use of the lemma in the natural argument for de-Poissonization, where we use the "add-one" bound (3.42) on the terms of the Poisson sum that are near the mean and we use the L^∞ bound for the others. Specifically, by Jensen's inequality and Hölder's inequality with $p = 4$, the fact that N has variance n gives us

$$\begin{aligned} |\phi(n) - EL_n^{\text{TSP}}| &\leq \sum_{k=0}^{\infty} |EL_k^{\text{TSP}} - EL_n^{\text{TSP}}| P(N = k) \\ &\leq \sum_{n/2 \leq k \leq 3n/2} c|n-k|n^{-1/2} P(N = k) \\ &\quad + cE\left(|N-n|^{1/2} \mathbf{1}(|N-n| \geq n/2) \right) \\ &\leq \frac{c}{\sqrt{n}} E|N-n| + cn^{1/4} P(|N-n| \geq n/2)^{3/4} \leq c. \end{aligned}$$

In other words, for Poisson N with mean n, we have

$$(3.43) \qquad EL_n^{\text{TSP}} - EL_N^{\text{TSP}} = O(1).$$

Now that we have a proper de-Poissonization inequality, we can immediately assert a proper analogue of (3.40) for problems of fixed-size n:

$$(3.44) \qquad \frac{EL_n^{\text{TSP}}}{\sqrt{n}} - \beta_{\text{TSP}}(2) = O(n^{-1/2}).$$

In the derivation of the basic rate result (3.40) for the Poisson TSP in $d = 2$, we took advantage of the fluke that for $d = 2$ the bounds in (3.38) and (3.39) just happened to match. We are not so lucky in general, so we need to develop tools that help us improve upon (3.39). There are now two general strategies for providing such inequalities. Perhaps the more natural of the two strategies is that introduced by Alexander (1994a). One idea central to Alexander's method is the that one can translate the usual partition Q_i, $1 \leq i \leq m^d$, of $[0,1]^d$ by a small amount and guarantee that the number of edges that meet $\cup \partial Q_i$ is much smaller than we supposed in the crude bounds applied in the derivation of (3.39).

This idea is then combined with a more careful execution of the patching strategy used to build suboptimal graphs for the subsquares.

The second strategy for improving upon (3.39) is due to Redmond and Yukich (1994). This method of "rooted duals" leads to a number of new ideas for the theory of subadditive Euclidean functionals, and it will be developed in detail in the next section.

3.8. Rooted duals and their applications.

For a subadditive Euclidean functional L on $[0,1]^d$, we can often find a closely related *superadditive* functional L^* that can help us profoundly with the analysis of L. A central feature of this helpful L^* is that it is a good approximation of L. In particular, we can require that L^* have expectation that satisfies

(3.45) $|EL(X_1, X_2, \ldots, X_n) - EL^*(X_1, X_2, \ldots, X_n)| \leq Cn^{(d-2)/d}$.

One technical point is that since the L^*'s that we can construct to satisfy (3.45) usually turn out not to be homogeneous or translation invariant, the superadditivity condition has to be phrased in a way that looks a little fussier than the familiar condition for geometric subadditivity. Still, the new condition carries the parallel intuition, and it is just as easy to use in applications.

Geometric superadditivity.

Let Q_i, $1 \leq i \leq m^d$, denote a partition of $[0,1]^d$ in to subcubes with side $1/m$, and let $\{q_i\}$ denote a set of vectors that translate Q_i back to the origin, that is, the $\{q_i\}$'s satisfy $Q_i - q_i = [0, 1/m]^d$. Also, for any $S = \{y_1, y_2, \ldots, y_k\} \subset \mathbb{R}^d$, we write mS for the set $\{my_1, my_2, \ldots, my_k\}$.

If there exists a constant $C_4 = C_4(L^*)$ such that for all integers $m \geq 1$, $n \geq 1$, and $\{x_1, x_2, \ldots, x_n\} \subset [0,1]^d$, we have

$$L^*(\{x_1, x_2, \ldots, x_n\})$$

(3.46) $$\geq \frac{1}{m} \sum_{i=1}^{m^d} L^*(m[(\{x_1, x_2, \ldots, x_n\} \cap Q_i) - q_i]) - C_4 m^{d-1},$$

then we say that L^* is *superadditive*.

If the functional L^* were translation invariant and homogeneous, then the sum in (3.46) would look just like the sum in our geometric subadditivity condition. The m's inside and outside the sum would be removed by homogeneity, and the q_i's would be removed by translation invariance. Evidently, one does not love (3.46) for its looks, but the inner attractiveness of (3.46) is easily brought out by the exploration of examples. The great fortune is that in all of the examples that matter to us most, there is a straightforward way to construct an associated L^* that satisfies both geometric superadditivity (3.46) and the approximation condition (3.45).

Rooted dual for the TSP.

Suppose that L is the usual TSP functional, so for any set $\{y_1, y_2, \ldots, y_s\} \subset \mathbb{R}^d$, the value $L(y_1, y_2, \ldots, y_s)$ is just the length of the shortest tour through the set $\{y_1, y_2, \ldots, y_s\}$. We will now introduce the *rooted dual* of L and see that it has the two basic properties of expectation approximation ((3.45)) and geometric superadditivity ((3.46)).

If $\mathcal{P} = \{F_1, F_2, \ldots, F_k\}$ is a collection of disjoint subsets of $F = \{x_1, x_2, \ldots, x_n\}$ whose union is F, we call $\mathcal{P}$ a partition of the set $F = \{x_1, x_2, \ldots, x_n\}$, and we let $\mathcal{P} = \mathcal{P}(F)$ denote the set of all partitions of F. The *rooted dual for the TSP* is defined for $x = \{x_1, x_2, \ldots, x_n\}$ by

$$L^*(x) = \min\left\{z = \sum_{i=1}^{k} L(F_i \cup \{a_i, b_i\}) : \{F_i\} \in \mathcal{P}(F), a_i, b_i \in \partial[0,1]^d\right\}.$$

In other words, L^* is the cost of the least expensive tour through the set of points $\{x_1, x_2, \ldots, x_n\}$, where the cost of travel within $[0,1]^d$ is the regular Euclidean distance cost but where travel along any path on the boundary of the cube $\partial[0,1]^d$ is free. If we have a partition $\{F_i\} \in \mathcal{P}(F)$ and a set of pairs of boundary points $\{a_i, b_i\}$, then the summand $L(F_i \cup \{a_i, b_i\})$ represents the cost of an optimum path that begins at a_i, travels through the points of F_i, and then goes to b_i. The interpretation of L^* is therefore the minimum sum of such terms over all choices of partitions and boundary pairs. Connections from one "loop" to the next are made by travel within the boundary, and thus they add nothing to the total cost.

The raison d'être of this definition is the transparent geometric superadditivity of L^*; we have (3.46) with $C_4 = 0$. To verify this, just note that if we take the optimal rooted path for the whole square, then the restriction to any subsquare is a set of loops that meet the boundary of the subsquare and that span the points of F that are in the subsquare. By suboptimality, we therefore find (3.46). To go the next step, the definition suggests that L^* should be quite close to L, but we will require a calculation to be sure that L^* is close enough to L to satisfy the quantitative relation (3.45).

To begin, we have one easy bound,

$$L^*(F) \leq L(F) + 1,$$

as one can see by taking the partition of F to be the trivial partition $\{F, \emptyset\}$. To get a bound that goes in the other direction, we need an estimate of the cardinality of the set $R \subset F$ of points that are rooted to the boundary; that is, R is the set of points that are joined by an edge to a boundary point. We consider one of the faces of $[0,1]^d$, say S. We let $R_S \subset R$ denote the set of points that are rooted to S, and we let $C(\epsilon, z)$ denote the cylinder in $[0,1]^d$ determined by the disk in S of radius ϵ around $z \in S$. The key observation is that there can be at most two points of R_S in the part of the set $C(\epsilon, z) \cap R_S$ that is at a distance greater than ϵ from S. The reason is simply that if there were three points of R_S, then there would be two from different loops, and these two could be joined with an edge and the rooting edges removed with a net cost savings of at least ϵ. Since S can be covered with $O(\epsilon^{-(d-1)})$ disks of radius ϵ, we have the bound

$$\text{card } R_S \leq \text{card } \{x \in F : \text{dist}(x, S) \leq \epsilon\} + c\epsilon^{-(d-1)},$$

or, by summing over all the $2d$ faces,

(3.47) $\qquad \text{card } R \leq \text{card } \{x \in F : \text{dist}(x, \partial[0,1]^d) \leq \epsilon\} + 2dc\epsilon^{-(d-1)}.$

If we take $\epsilon = n^{-1/d}$, we find $E(\text{card } R) = O(n^{(d-1)/d})$. If we let $R' \subset \partial[0,1]^d$ denote the set of points that are joined to R, then card $R' = $ card R, and by combining the rooted tour with an optimal tour of R', we get a connected graph that contains a tour of F, so $L(F) \leq L^*(F) + L(R') + 1$. Since any k points of $\partial[0,1]^d$ are on a tour of length no greater that $ck^{(d-2)/(d-1)}$, Jensen's inequality implies $EL(R') = O(n^{(d-2)/d})$ so that we have proved the expectation-approximation inequality (3.45) for the TSP functional.

Rooted dual for the MST.

The invention of a rooted dual L^* for the MST requires only the most modest variation on the previous example. After a moment's consideration, one discovers that an appropriate L^* is given by

$$L^*(x) = \min \left\{ \sum_{i=1}^k L(F_i \cup \{a_i\}) : \{F_i\} \in \mathcal{P}(F), a_i \in \partial[0,1]^d \right\}.$$

Thus we see that for the MST, we only need to have one edge that connects F to the boundary. Again, we see that L^* is superadditive with C_4 equal to zero. The proof of the expectation approximation can be handled just as before, even with modest improvements in some of the bounds. In particular, this time the "key observation" can be modified to say that there can be at most one point of R_S in the part of the set $C(\epsilon, z) \cap R_S$ that is at a distance greater than ϵ from S.

Rooted dual for the minimal matching.

Here there is a modest novelty in the definition of L^* since in this case we do not insist on having a point of F matched to the boundary. Formally, we let

$$L^*(F) = \min\{L(F \cup G) : G \subset \partial[0,1]^d\}.$$

This means nothing more than that in our calculation of L^*, we have a minimal matching where we can match any points that we like to the boundary, but we are not forced to match any points to the boundary. Superadditivity (with $C_4 = 0$) is immediate, and the bound on the number of rooted points can be achieved just as in the case of the MST.

Rate results for the means.

The rooted duals for the TSP, MST, and minimal matching give us an easy road to a rate result for the convergence of the mean values. Specifically, if $\{X_i : 1 \leq i < \infty\}$ are independent and uniformly distributed on $[0,1]^d$, then for any of these functionals, we have for all $d \geq 2$ that

$$EL_n = \beta_L(d) n^{(d-1)/d} + O\left(n^{(d-2)/d}\right).$$

The proof of this result is straightforward given the techniques that have been developed in the earlier sections. The first step is to consider the corresponding problem where we have a Poisson sample size. From subadditivity and the existence of the superadditive dual that satisfies $EL_n = EL_n^* + O(n^{(d-2)/d})$, we find that if we set $\phi(n) \equiv EL_N$, where N has the Poisson distribution with mean n, then $\phi(n)$ will satisfy

$$|\phi(2^d n) - 2^{d-1}\phi(n)| \leq cn^{(d-2)/d},$$

or, dividing by $2^{d-1} n^{(d-1)/d}$ and taking advantage of generic constants, we find

$$\left| \frac{\phi(2^d n)}{(2^d n)^{(d-1)/d}} - \frac{\phi(n)}{n^{(d-1)/d}} \right| \leq cn^{-1/d}.$$

Generally, when we replace n by $2^{kd} n$, we find

$$\left| \frac{\phi(2^{(k+1)d} n)}{(2^{(k+1)d} n)^{(d-1)/d}} - \frac{\phi(2^{kd} n)}{(2^{kd} n)^{(d-1)/d}} \right| \leq c(2^k n)^{-1/d};$$

so we sum to find

$$\left| \beta_L(d) - \frac{\phi(n)}{n^{(d-1)/d}} \right| \leq cn^{-1/d},$$

where $\beta_L(d) = \inf \phi(n)/n^{(d-1)/d}$. This is precisely the required rate result in the Poisson case.

Passage from uniform to general densities.

If L is a subadditive Euclidean functional for which there is a process L^* that satisfies functional smoothness in the sense of (3.26), closeness of expectations expressed by (3.45), and geometric superadditivity (3.46), one can provide an extension to general densities in a way that parallels the first three steps of our proof of the BHH theorem. In particular, this means that for L equal to the MST and minimum-matching problems, one has a constant $\beta_L(d)$ such that for any sequence of independent random variables $\{X_i\}$ with density f with compact support, we have

(3.48) $\quad L(X_1, X_2, \ldots, X_n)/n^{(d-1)/d} \to \beta_L(d) \int_{\mathbb{R}^d} f(x)^{(d-1)/d} \, dx \quad$ a.s.

3.9. Lower bounds and best possibilities.

The development of rate results would be never ending if there were no results that told us that tell us when we can do no better. For the TSP, MST, minimal-matching, and many other problems of geometric combinatorial optimization, we now know that

$$EL_n = \beta_L(d) n^{(d-1)/d} + O(n^{(d-2)/d}),$$

so the inevitable question is whether these results are the best possible. The state of the theory is not 100% complete, but we do have strong indications that these rate results cannot be improved.

To give one example of what is known, we note that Rhee (1994) has proved that there is a constant $c > 0$ such that for Poisson N with mean n, one has

$$(3.49) \qquad \frac{EL_N^{\text{TSP}}}{\sqrt{n}} > \beta_{\text{TSP}}(2) + cn^{-1/2},$$

giving us the very nice fact that the result (3.40) is the best possible.

The idea of the proof of (3.49) is that in the partition of the unit square into four subcubes Q_i, $1 \leq i \leq 4$, the optimal tour over the whole sample $\{X_1, X_2, \ldots, X_N\}$ can (with high probability) achieve meaningful savings over the sum of the four terms $L^{\text{TSP}}(Q_i \cap \{X_1, X_2, \ldots, X_N\})$. The details of this fact for the TSP are too taxing to recall here, but for the minimal spanning tree, one has a comparable result with far fewer geometric details. We will therefore illustrate the technique for obtaining lower bounds like (3.49) by consideration of the minimal-spanning-tree problem.

Lower-bound method and the MST.

For the minimal spanning tree, the key relationship that leads us to a lower bound like (3.49) can be put into a lemma.

LEMMA 3.9.1 (Jaillet (1993)). *If N_n denotes a Poisson random variable with mean n, and we write $L^{\text{MST}}(N_n) = L^{\text{MST}}(X_1, X_2, \ldots, X_{N_n})$ for the length of the minimal spanning tree of an independent uniform sample of size N_n from the unit square, then there is a constant $c > 0$ such that*

$$(3.50) \qquad EL^{\text{MST}}(N_{4n}) \leq 2EL^{\text{MST}}(N_n) - c.$$

Before proving the lemma, we should note how it relates to the work we have reviewed for the TSP. We will not give the argument right now, but one can show that the MST satisfies a two-sided bound inequality just like (3.38)–(3.39); so one can repeat verbatim the proof that was given for the TSP to establish that

$$(3.51) \qquad \frac{EL_N^{\text{MST}}}{\sqrt{n}} - \beta_{\text{MST}}(2) = O(n^{-1/2}).$$

Thus a lower bound of the form

$$(3.52) \qquad \frac{EL_N^{\text{MST}}}{\sqrt{n}} > \beta_{\text{MST}}(2) + cn^{-1/2} \quad \text{with } c > 0$$

would be the perfectly parallel result to (3.49). This is just what we will extract from Lemma 3.9.1.

To see how Lemma 3.9.1 leads to (3.52), we note that if we begin with $4^m n$ in place of $4n$ in (3.50) and apply Lemma 3.9.1 recursively, we have

$$EL^{\text{MST}}(N_{4^m n}) \leq 2^m EL^{\text{MST}}(N_n) - c(2^m - 1).$$

So if we divide by $\sqrt{4^m n}$ and let $m \to \infty$, we find $\beta_{\text{MST}}(2) + c \leq EL^{\text{MST}}(N_n)/\sqrt{n}$, or, in other words, (3.51) is the best possible.

So now the issue is the proof of Lemma 3.9.1. We let $S \equiv \{x_1, x_2, \ldots, x_n\} \subset [0,1]^2$ and form the minimal spanning trees T_i for each $S \cap Q_i$, $1 \leq i \leq 4$. We then choose the point $x \in S$ that is nearest to the midpoint of $[0,1]^2$, and for each of the trees T_i such that $x \notin T_i$, we add an edge between x and the nearest point of T_i to x. This gives us a spanning tree T of S. Now the key idea is that we can do a little tree surgery on T to get a new spanning tree T' that is shorter than T. The source of the potential savings between T and T' is that there may be points that are near the interior boundaries of the Q_i that might be better matched across the boundary than to the tree $T_i \subset Q_i$.

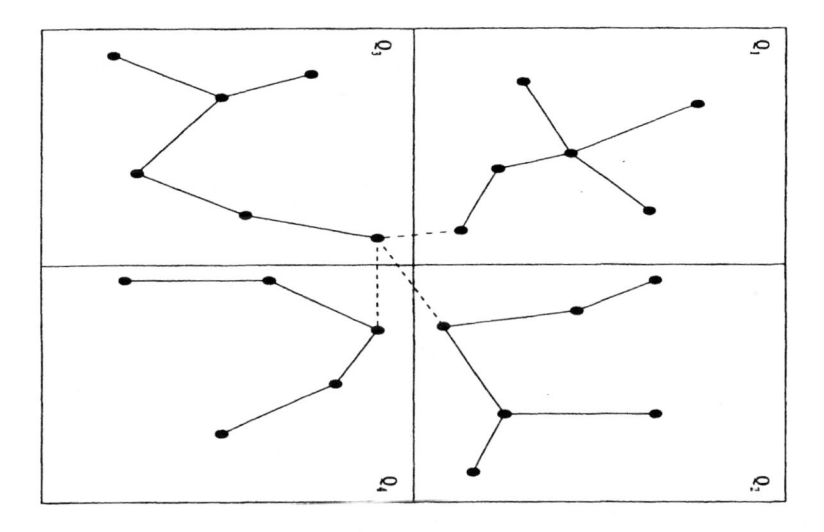

FIG. 3.1. *A spanning tree that can be improved.*

We begin by considering a little geometry that is indicated in Figure 3.2. Suppose that we have a disk D of radius $4r$ that is contained in $Q_1 \cap Q_2$ and whose center is on the boundary B between Q_1 and Q_2. Suppose further that $D' \subset D$ is a disk of radius r with the same center as D. Now suppose that we choose two points at random in D, and we consider the event that both of these points actually fall into D' with exactly one of these falling into $Q_1 \cap D'$ and the other one falling into $Q_2 \cap D'$. The blandest conceivable fact about this event is that is has probability $p > 0$ that does not depend on r.

Now we place, say, $\lfloor n^{\frac{1}{2}} \rfloor$ disks $D_i \subset Q_1 \cup Q_2$ of radius $4r = 5^{-1} n^{-\frac{1}{2}}$ with their centers along the boundary line B. We also consider disks D'_i of radius r and the same center as D_i. If we have a Poisson sample of size N_{4n} with mean $4n$, then for any given i, the probability that D_i contains exactly two points is bounded below by a $\tau > 0$, where τ is a constant that does not depend upon n.

Next, we look at a sample of uniformly distributed random variables $X_1, X_2, \ldots, X_{N_{4n}}$ and we construct the spanning tree T_i for $\{X_1, X_2, \ldots, X_{N_{4n}}\} \cap Q_i$ for $1 \leq i \leq 4$. We join these trees together as described above to form a spanning tree T of $\{X_1, X_2, \ldots, X_{N_{4n}}\}$. Now consider the set I of all $i \in [1, n^{\frac{1}{2}}]$ such that

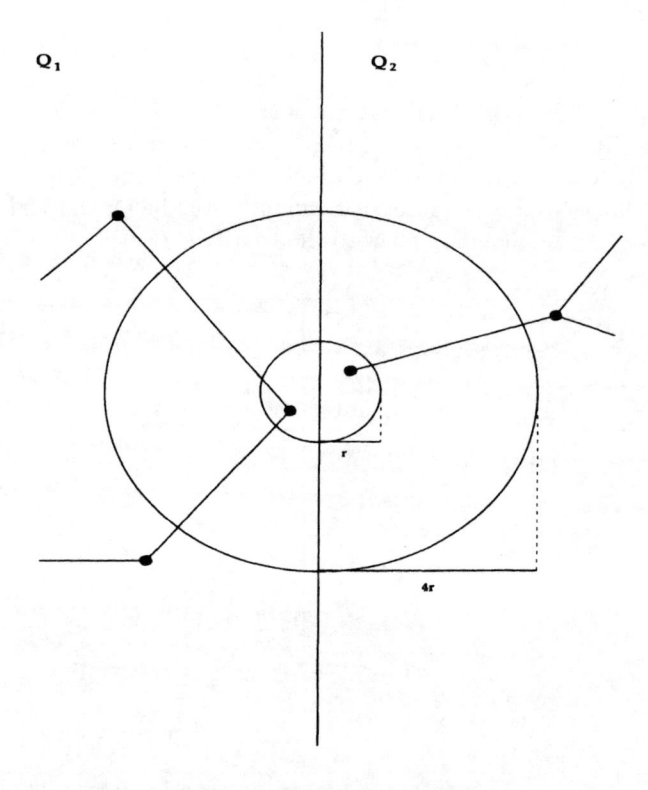

FIG. 3.2. *A circle offering an opportunity.*

- D_i contains exactly two points of $\{X_1, X_2, \ldots, X_{N_{4n}}\}$,

- $Q_1 \cap D'_i$ contains exactly one point, say Y_i, and

- $Q_2 \cap D'_i$ contains one exactly one point, say Y'_i.

We also observe that I is not too small; specifically, $EI \geq p\tau \lfloor n^{\frac{1}{2}} \rfloor$, where p and τ do not depend on n.

For each $i \in I$, we also let A_i denote the set of edges of $T_1 \cup T_2$ that are incident to Y_i or Y'_i. We note that all of the edges of A_i have length at least as large as $3r$ since they have to stretch from the inside of a small disk to the outside of a big disk.

If we now add the edge (Y_i, Y'_i) edge to T, we create a cycle, and one of the edges of A_i, say e, must be on that cycle. We then delete e to get back to a spanning tree. The cost of the added edge (Y_i, Y'_i) is at most $2r$ since both are elements of D'_i, and as we observed earlier, the cost of the deleted edge is at least $3r$. This gives us a net savings of r for each element of the set I. We have $4r = 5^{-1}n^{-\frac{1}{2}}$ and $EI \geq p\tau \lfloor n^{\frac{1}{2}} \rfloor$, so $rEI \geq c_0 > 0$ for a constant c_0 that does not depend on n, so

$$EL^{\mathrm{MST}}(N_{4n}) \leq EL^{\mathrm{MST}}(N_n) + c_1 n^{-1/2} - c_0.$$

We therefore find that for all $n \geq n_0 \equiv 4c_1/c_0^2$, we have the inequality of the lemma with $c = c_0/2$, and the proof of the lemma is complete.

Issue of fixed n.

Since the minimal spanning tree has exactly the same "add-one" inequality $|L_n^{\text{MST}} - L_{n+1}^{\text{MST}}| \leq c\sqrt{n}$ that we used in our de-Poissonization argument for the rate result for the TSP, we can argue just as in (3.43) to find $EL_{N_n}^{\text{MST}} - EL_n^{\text{MST}} = O(1)$, from which we immediately extract the $d = 2$ rate result for the MST:

$$\beta_{\text{MST}}(2) + c \leq EL_n^{\text{MST}}/\sqrt{n}. \tag{3.53}$$

Are the best possible fixed-sample-size rate results those given by (3.44) for the TSP and by (3.53) the MST? We have established as much for the Poisson-sample-size problems, but this time our techniques for de-Poissonization come up short. One surely suspects that these results and their d-dimensional analogues cannot be improved, but so far we do not know a proof.

Rhee (1994) provides a discussion of this problem and puts forward three closely related conjectures that underscore the difficulties that one might encounter. Rhee's third conjecture is perhaps the most telling:

There is a constant ψ such that for all $\epsilon > 0$ and all $n \geq n(\epsilon)$, we have for $F = \{X_1, X_2, \ldots, X_N\}$, where N is Poisson with parameter n, that

$$P\left(\left| L^{\text{TSP}}(F) - \sum_{1 \leq i \leq 4} L^{\text{TSP}}(F \cap Q_i) \right| \geq \epsilon \right) \leq \epsilon. \tag{3.54}$$

This conjecture may look innocent, and it may be, but if (3.54) could be proved, then the central limit theorem (CLT) for $L^{\text{TSP}}(F)$ would follow. All one would have to do is iterate the inequality until one has an inequality that gives a good bound on the difference between $L^{\text{TSP}}(F)$ and a sum of 2^k independent terms. One could then invoke the usual central limit theorem on the sum. In fact, even a little less than Rhee's conjecture would suffice for the CLT—one can permit ψ to depend on n rather than be a constant.

3.10. Additional remarks.

The notes below provide some additional details on the evolution of the results that have been covered in this chapter as well as pointers toward work that has important connections to the material that we have covered.

1. Theorem 3.1.1 was proved in Steele (1981a) with the additional hypothesis that $\text{Var } L(X_1, X_2, \ldots, X_n) < \infty$ for $\{X_i\}$ independent with the uniform distribution on $[0, 1]^d$. When Rhee (1993a) established Lemma 3.3.1, it became evident that the variance assumption of Steele (1981a) could be removed using a simplified version of Rhee's argument.

2. The invitation after (3.17) to engage in some modest soul searching is motivated by the historical development of the method that was used to prove Theorem 3.1.1. Kingman (1973) had used Poisson embedding and his own subadditive

ergodic theorem to show that in the longest-increasing-subsequence problem, one has convergence in probability of $I_n/\sqrt{n}$, but because of the way the Poisson embedding related to the subadditive ergodic theorem in Kingman's proof, the inference could not be strengthened to an almost sure theorem.

Kesten (1973) noticed that the almost sure version of the limit theorem could be saved if one avoided the subadditive ergodic theorem and used a more direct technique motivated by Richardson (1973). The essence of Richardson's technique and Kesten's corresponding limit theorem is that one works toward an inequality like (3.14). Since such a sum just depends upon the probability law (and not the probability space), the nuance that had one stopped at a convergence in probability law could be overcome. Another interesting feature of Richardson's technique is that it reinforces the observation that if one has monotonicity then one can work with very sparse subsequences and still extract convergence results.

3. The limit theory for the matching problem was first discussed in Papadimitriou (1978b), where a partitioning algorithm for the Euclidean matching problem is developed. Papadimitriou (1978b) also provides some information about the limiting constant in $d = 2$.

4. Mézard and Parisi (1988) offer an approach to the Euclidean matching problem that is completely different from those that have been reviewed here. Their analysis is based on the replica method of statistical mechanics, and—though fascinating—the analysis would seem to stand at a considerable distance from the parts of statistical mechanics that can now be made rigorous.

5. The method of rooted duals has been applied in Yukich (1996a) to obtain results for the worst-case behavior of the TSP, MST, and other functionals. Also, in Yukich (1996b), a connection has been made between higher-dimensional subadditive ergodic theorems and subadditive Euclidean functionals. This approach yields limit theorems that can rely on stationarity of the underlying process rather than independence.

CHAPTER 4

Probability in Greedy Algorithms and Linear Programming

In this chapter, the assignment problem serves as our leading example for the analysis of greedy algorithms, matching strategies, and—most importantly—the interplay of probability theory and linear programming. The main theoretical result that is proved is the Dyer–Frieze–McDiarmid inequality, which gives a remarkably general bound on the expected value of the objective function of a linear program with random cost.

4.1. Assignment problem.

Consider the task of choosing an assignment of n jobs to n machines in order to minimize the total cost of performing the n jobs. The basic input for the problem is an $n \times n$ matrix (c_{ij}), where c_{ij} is viewed as the cost of performing job i on machine j, and the *assignment problem* is to determine a permutation π that solves

$$(4.1) \qquad A_n = \min_{\pi} \sum_{i=1}^{n} c_{i\pi(i)}.$$

No doubt the simplest stochastic model for the assignment problem is given by considering the c_{ij} to be independent random variables with the uniform distribution on $[0, 1]$. This model is apparently quite simple, but after some analysis one finds that it possesses a considerable richness. Exact or approximate solutions of the assignment problem can be obtained by many different methods, and each of these methods seems to have something new to say about the stochastic model.

Local greedy assignment.

Greedy algorithms are among the most studied methods in combinatorial optimization, and there are important problems like the determination of a minimal spanning tree where the natural greedy algorithms are indeed optimal. In the case of the assignment problem, the greedy approach does not lead to an optimal solution, and the extent of this failure is one of the reasons that analysis of the assignment problem is particularly instructive.

Consider the cost H_n of the heuristic assignment given by successively examining each job i, $1 \leq i \leq n$, and making an assignment to the free machine for which c_{ij} is minimal, i.e., $\pi(i)$ is chosen so that $c_{i\pi(i)} = \min\{c_{ij} : j \neq \sigma(k)$ for all $k < i\}$. Under the hypothesis that the $\{c_{ij}\}$'s are independent and uniformly distributed on $[0,1]$, the ith assignment equals the minimum of $n-i+1$ independent uniformly distributed random variables, so the expected cost of the ith assignment is exactly $1/(n-i+2)$. As a consequence, we have

$$EH_n = \sum_{1 \leq i \leq n} 1/(n-i+2) \sim \log n.$$

This analysis is certainly easy, but shortly we will find that we can make less costly assignments. For the moment, we should just note that by a similar computation we can check that $EH_n^2 = O(\log^2 n)$. This fact will turn out to be useful later in our analysis of a more refined heuristic method.

Global greedy assignment.

A less naïve heuristic for the assignment problem chooses matches successively out of the set of all possible matches by choosing the pair i and j for which c_{ij} is the least cost among the unmatched pairs. This method looks promising since the expected cost of the first match is just $1/(n^2+1)$ rather than the $1/(n+1)$ of the local greedy assignment. Still, this early success soon falters, and the global greedy heuristic again leads to an expected total cost of order $\log n$.

To see why this is so requires only a little work. To develop a recursion via a first-step analysis, we let $a_n(t)$ denote the expected cost of the assignment produced by the global greedy heuristic under the assumption that the costs c_{ij} of the n^2 job-processor pairs are independent and uniformly distributed on the restricted interval $[t,1]$. To solve the original problem, we therefore need to determine $a_n(0)$.

If $m_k(t,u)$ denotes the density of the minimum of k^2 random variables with the uniform distribution in $[t,1]$, we easily see from the definition of the global greedy heuristic that we have the recursion

$$a_n(t) = \int_t^1 \{u + a_{n-1}(u)\} m_n(t,u)\,du.$$

What makes this easy to solve is the observation that by scaling we have $a_n(t) = nt + (1-t)a_n(0)$. Thus if we write a_n for $a_n(0)$ and substitute into the integral recursion, we find

$$a_n = \int_0^1 u m_n(0,u)\,du + \int_0^1 \{(n-1)u + (1-u)a_{n-1}\} m_n(0,u)\,du$$
$$= \frac{n}{(1+n^2)} + a_{n-1}\frac{n^2}{(1+n^2)}.$$

This recursion easily shows that a_n is exactly of order $\log n$, a result that goes back to Kurtzburg (1962), where several heuristic methods for the assignment problem were first studied.

Assignments via matching theory.

The greedy methods turn out to be wide of the mark, since one can show that for the uniform model the expected minimal cost assignment of the n-machine,

n-job assignment problem actually bounded independently of n. More precisely, Walkup (1979) showed the cost A_n of minimal assignment satisfies $EA_n \leq 3 + o(1)$. There are even sharper bounds that we will investigate at some length, but we will still want to review Walkup's approach since it has several ideas that should be useful in other problems.

One compellingly natural approach to the stochastic assignment problem is to look for an assignment that only uses edges with small costs. Ideally, we might consider the bipartite graph such that the edge (i,j) is in the edge set E if and only if $c_{ij} \leq \alpha/n$. If such a bipartite graph were to have a perfect matching, then the cost of the assignment problem would be bounded by α. Unfortunately, this method does not yield the constant bound that we seek since, among other troubles, the expected number of *isolated* vertices is $2n(1 - \alpha/n)^n \sim 2ne^{-\alpha}$. Further consideration of this fact reveals that we cannot expect to find a perfect matching with all edge lengths less than α/n unless we take $\alpha = \Omega(\log n)$, in which case we would end up with a bound on EA_n that is no better than the bound we found by the greedy methods.

The key to doing better is to find some way to make sure that every vertex has at least a few edges, and the idea of k-out multigraphs turns out to serve us well here. To define the *random two-out bipartite multigraph*, we first let $\{V_1, V_2\}$ with card $V_i = n$ be a bipartition for a vertex set with $2n$ vertices. Next, for each $v \in V_1$, we choose at random two elements w and w' from V_2, and we take the edges (v,w) and (v,w') as elements of our edge set. Similarly, for each $v \in V_2$, we choose at random two elements of w and w' from V_1, and we also add the corresponding edges to our edge set. Since we may end up with some edge added twice to our edge set, we have a multigraph rather than a graph; but this distinction turns out to be inconsequential. The real point is that now with high probability, we have a perfect matching because the troubles related to isolated points have been handled. The critical fact is expressed in the following theorem from Walkup (1981).

THEOREM 4.1.1. *Let G be a random two-out bipartite multigraph with bipartition $\{V_1, V_2\}$ with card $V_i = n$. The probability that G fails to contain a perfect matching is bounded by $5/n$.*

There is a now a nice trick that lets us connect the assignment problem to the theory of random two-out bipartite multigraphs. For each edge cost c_{ij}, we introduce a pair of edge weights $c_{ij}^{(1)}$ and $c_{ij}^{(2)}$ as follows:

(i) we choose a number u_{ij} from the interval $[c_{ij}, 1]$ according to the distribution

$$F(x) = \{(1 - c_{ij})^{1/2} - (1 - x)^{1/2}\}/\sqrt{1 - c_{ij}},$$

and

(ii) with probability $\frac{1}{2}$ we set $c_{ij}^{(1)} = u_{ij}$ and $c_{ij}^{(2)} = u_{ij}$. Otherwise, we do just the opposite, that is, we set $c_{ij}^{(2)} = u_{ij}$ and $c_{ij}^{(1)} = c_{ij}$.

By the construction, we always have the identity

$$\min\left\{c_{ij}^{(1)}, c_{ij}^{(2)}\right\} = c_{ij},$$

and we have considerable information in addition. First, by direct calculation, one can check that $c_{ij}^{(1)}$ and $c_{ij}^{(2)}$ both have the unconditional distribution on $[0,1]$ given by

$$F(x) = 1 - (1-x)^{1/2},$$

and, more importantly, the direct computation of the unconditional joint distribution shows that $c_{ij}^{(1)}$ and $c_{ij}^{(2)}$ are independent of each other.

The collection of $4n$ independent random variables $\{c_{ij}^{(1)}, c_{ij}^{(2)} : 1 \leq i, j \leq n\}$ provides just the tool that we need to define a bipartite two-out multigraph that can be used to provide bounds for the assignment problem. The idea is to take the two edges out of each $i \in V_1$ that correspond to the two smallest values of $c_{ij}^{(1)}$ with $1 \leq j \leq n$ and to take an analogous pair of edges out of each $j \in V_2$. Formally, for each $i \in V_1$, we take the edges (i,s) and (i,t), where $c_i^{(1)}(1) \equiv c_{is}^{(1)}$ and $c_i^{(1)}(2) \equiv c_{it}^{(1)}$ are, respectively, the smallest and the second smallest elements of $\{c_{ij}^{(1)} : 1 \leq i,j \leq n\}$. Similarly, for each $j \in V_2$ we take the edges (s,j) and (t,j) where $c_j^{(2)}(1)$ and $c_j^{(2)}(2)$ are the respectively the smallest and second smallest elements of $\{c_{ij}^{(2)} : 1 \leq i \leq n\}$. By the independence and identical distribution of the variables in the set $\{c_{ij}^{(1)}, c_{ij}^{(2)} : 1 \leq i \leq j \leq n\}$, we see that the G defined by this process is precisely a random two-out multigraph.

The other fact that we need to take into account is that we have the expectations for the smallest values

$$Ec_i^{(1)}(1) = Ec_j^{(2)}(1) = \frac{2}{n}$$

and the second smallest values

$$Ec_i^{(1)}(2) = Ec_j^{(2)}(2) = \frac{4}{n}.$$

Also, if we let O_i be the set of out-edges of the vertex $i \in V_1$ and let O_j' be the set of out-edges of the vertex $j \in V_2$, then

$$E\left(c_{ij}^{(1)} | (i,j) \in O_i\right) = \frac{1}{2} E\left(c_{ij}^{(1)} | c_{ij}^{(1)} = c_i^{(1)}(1)\right) + \frac{1}{2} E\left(c_{ij}^{(1)} | c_{ij}^{(1)} = c_i^{(2)}(1)\right) = \frac{3}{n},$$

and in just the same way, we have

$$E\left(c_{ij}^{(2)} | (i,j) \in O_j'\right) = \frac{3}{n}.$$

Naturally, these identities explain the source of the magical constant 3 in Walkup's inequality.

We have everything we need to calculate the matching bound for EA_n, so we only need to assemble the pieces. First, we observe that for *any* matching M, we have

(4.2) $$\sum_{(i,j) \in M} c_{ij} = \sum_{(i,j) \in M} \min\left\{c_{ij}^{(1)}, c_{ij}^{(2)}\right\}$$
$$\leq \sum_{(i,j) \in M} \frac{1}{2}\left\{c_{ij}^{(1)} + c_{ij}^{(2)}\right\}.$$

We now let B denote the event that G contains a matching, and we let M_k for $k = 1, 2, \ldots, T$ be a list of the possible matchings of a generic n–n bipartite graph; we will use these matchings to build a partition of B. First, we let B_1 be the event that M_1 is a matching in G, and subsequently we let B_k be the event that M_k is a matching in G but no M_j with $j < k$ is a matching of G. The B_k's then form a finite sequence of disjoint events such that $B = \cup B_k$. By the construction of G, the presence of an edge (i,j) in G does not depend on the cost of the edge except to the extent that the edge is in either O_i or O'_j. This tells us that once we know that $(i,j) \in O_i$, there is no additional information about $c_{ij}^{(1)}$ to be gained from knowing $c_{ij}^{(1)} \in M_k$. In particular, we have

$$(4.3) \qquad E\left(c_{ij}^{(1)} | (i,j) \in O_i \text{ and } B_k\right) = E\left(c_{ij}^{(1)} | (i,j) \in O_i\right) = \frac{3}{n}$$

and

$$(4.4) \qquad E\left(c_{ij}^{(2)} | (i,j) \in O'_j \text{ and } B_k\right) = E\left(c_{ij}^{(2)} | (i,j) \in O_i\right) = \frac{3}{n}.$$

Next, let C denote the complement of the set B. If H_n is the cost of the local greedy heuristic assignment discussed at the beginning of the chapter, then together with the bound (4.2), we have

$$A_n \leq 1(C) H_n + \sum_k 1(B_k) \sum_{(i,j) \in M_k} \frac{1}{2}\left\{c_{ij}^{(1)} + c_{ij}^{(2)}\right\}.$$

Now Theorem 4.1.1 tells us that $P(C) \leq 5/n$, so if we use (4.3) and (4.4) together with our earlier observation that $EH_n^2 = O(\log^2 n)$, we find

$$EA_n \leq 3P(B) + P(C)^{1/2} \|H_n\|_2 = 3 + O(n^{-1} \log n).$$

This is precisely what we wished to prove.

Linear programming connection.

The theory of linear programming deals with problems that can be formulated in terms of the minimization of a linear function subject to constraints. The generic linear program in standard form imposes linear *equality constraints* together with just positivity constraints on the underlying variables:

$$\begin{aligned} \text{minimize} \quad & \sum_{j=1}^n c_j x_j \\ \text{subject to} \quad & \sum_{j=1}^n a_{ij} x_j = b_i, \quad i = 1, 2, \ldots, m, \\ & x_j \geq 0, \qquad j = 1, 2, \ldots, n. \end{aligned}$$

One of the most basic themes of combinatorial optimization is that many of its fundamental problems may be transformed into equivalent linear programs. Sometimes this transformation is straightforward, but in other cases— such as in the traveling-salesman problem or minimal matching—the development of appropriate transformations requires considerable imagination. Numerous benefits accrue from expressing combinatorial problems as linear programs because the translation makes available powerful tools such as the duality theory of linear programming and the theory of the simplex algorithm. Moreover, a linear-programming representation gives one access to many approximations and heuristics that have been developed for integer and linear programming, such as the method of Lagrangian relaxation and the method of cutting planes.

The main point of this chapter is to illustrate how probability theory can be used within the theory of linear programming to provide insight into problems of combinatorial optimization. The result that carries the bulk of this message is the inequality of Dyer, Frieze, and McDiarmid (1986), which provides a bound on the expectation of the value of the solution of a linear program with random variables as cost coefficients. This bound is provided in terms of the expected value of the costs and the value of a nonrandom feasible solution to the linear program. This extraordinary result has all the charms one can hope for in a mathematical result. It comes naturally out of an important theory—the simplex method; it is simple and easily applied; and, finally, it is reasonably sharp in examples of significance.

A first look at (4.1) may make one uncertain about the connection with linear programming since the problem of minimization over the set of permutations has the distinct flavor of combinatorial optimization. A first pass at expressing (4.1) as a linear program leads us to investigate the problem

(4.5)
$$\begin{aligned}
\text{minimize} \quad & \sum_{i,j} c_{ij} x_{ij} \\
\text{subject to} \quad & \sum_{j=1}^{n} x_{ij} = 1, \quad i = 1, 2, \ldots, n, \\
& \sum_{i=1}^{n} x_{ij} = 1, \quad j = 1, 2, \ldots, n, \\
\text{and} \quad & x_{ij} \geq 0.
\end{aligned}$$

If we further require the x_{ij} to be integers, problem (4.5) is clearly equivalent to problem (4.1), but if we do not explicitly impose an integrality constraint, we might reasonably doubt that these two problems are equivalent. Still, a fortunate coincidence takes place, and, as we will soon check, we do not need to impose any extra conditions on the linear program (4.5) for it to be equivalent to (4.1). This fact is one of a suite of *integrality theorems* that help linear programming serve combinatorial objectives. The result that we need for the assignment problem will easily emerge from one of the simplest and oldest of the integrality theorems that we develop in a later section. First, we need to develop some tools that are well known to optimizers but little used by probabilists.

4.2. Simplex method for theoreticians.

The general linear-programming problem is traditionally written in matrix form as

(4.6)
$$\begin{aligned}
\text{minimize} \quad & z = \mathbf{cx} \\
\text{subject to} \quad & \mathbf{Ax} = \mathbf{b}, \\
& \mathbf{x} \geq 0.
\end{aligned}$$

PROBABILITY IN GREEDY ALGORITHMS AND LINEAR PROGRAMMING   83

Here, of course, $\mathbf{A}$ is an $m \times n$ matrix, $\mathbf{x}$ is an n-dimensional column vector, $\mathbf{b}$ is an m-dimensional column vector, and $\mathbf{c}$ is an n-dimensional row vector. We will often find it useful to write $\mathbf{A} = [\mathbf{a}_1, \mathbf{a}_2, \ldots, \mathbf{a}_n]$, where $\mathbf{a}_j$ denotes the jth column of $\mathbf{A}$.

In a nutshell, the idea of the simplex method is that one can rewrite problem (4.6) in an equivalent form that either suggests further reformulations or makes the solution obvious. The central concept behind this sequence of reformulation is the notion of a *basis* B which is a subset of $\{1, 2, \ldots, n\}$ of cardinality m with the property that the corresponding columns of $\mathbf{A}$ are linearly independent. The variables x_i such that $i \in B$ are called the *basic variables*. The complement of B is denoted by N, and the x_i's such that $i \in N$ are called *nonbasic variables*.

The first step of our reformulation takes us from the equation $\mathbf{Ax} = \mathbf{b}$ to one that expresses the basic variables as a linear combination of the nonbasic variables and the constraint variable $\mathbf{b}$. Thus if we separate $\mathbf{x}$ and $\mathbf{A}$ into components consisting of the basic and nonbasic variables and collect these on different sides of the equation, we can write

(4.7) $$\mathbf{A}_B \mathbf{x}_B = \mathbf{b} - \mathbf{A}_N \mathbf{x}_N.$$

For example, if we have

$$\mathbf{A} = \begin{bmatrix} 1 & 1 & 2 & 1 & 5 \\ 1 & 1 & 2 & 0 & 1 \\ 2 & 1 & 3 & 1 & 0 \end{bmatrix}, \quad \mathbf{b} = \begin{bmatrix} 39 \\ 8 \\ 5 \end{bmatrix}, \quad \mathbf{x} = \begin{bmatrix} x_1 \\ x_2 \\ x_3 \\ x_4 \\ x_5 \end{bmatrix}$$

and $\mathbf{B} = \{x_2, x_3, x_5\}$, then the decomposition (4.7) is given by

$$\mathbf{A}_B = \begin{bmatrix} 1 & 2 & 5 \\ 1 & 2 & 1 \\ 1 & 3 & 0 \end{bmatrix}, \quad \mathbf{A}_N = \begin{bmatrix} 1 & 1 \\ 1 & 0 \\ 2 & 1 \end{bmatrix}, \quad \mathbf{x}_B = \begin{bmatrix} x_2 \\ x_3 \\ x_5 \end{bmatrix}, \quad \mathbf{x}_N = \begin{bmatrix} x_1 \\ x_5 \end{bmatrix}.$$

The first step of the reformulation is completed by using (4.7) to provide an expression for $\mathbf{x}_B$:

(4.8) $$\mathbf{x}_B = \mathbf{A}_B^{-1} \mathbf{b} - \mathbf{A}_B^{-1} \mathbf{A}_N \mathbf{x}_N.$$

We should note that the inverse of $\mathbf{A}_B$ exists because our definition of the basis requires the linear independence of the columns of the square matrix $\mathbf{A}_B$.

Next, we express the objective function $z = \mathbf{cx}$ in a form that reflects the division into basic and nonbasic variables. Certainly, we can write $z = \mathbf{c}_B \mathbf{x}_B + \mathbf{c}_N \mathbf{x}_N$, but by using (4.8) we can clear away the basic variables entirely and write the original problem (4.6) in the very useful *dictionary* form:

(4.9) $$\begin{aligned} \mathbf{x}_B &= \mathbf{A}_B^{-1} \mathbf{b} - \mathbf{A}_B^{-1} \mathbf{A}_N \mathbf{x}_N, \\ z &= \mathbf{c}_B \mathbf{A}_B^{-1} \mathbf{b} + (\mathbf{c}_N - \mathbf{c}_B \mathbf{A}_B^{-1} \mathbf{A}_N) \mathbf{x}_N. \end{aligned}$$

We are now in position to see the point of this maneuvering. Suppose we are lucky and all the coefficients of the nonbasic variables in the formula for z in (4.9) are nonnegative, i.e., suppose that

(4.10) $$c_j \geq \mathbf{c}_B \mathbf{A}_B^{-1} \mathbf{a}_j \quad \text{for all } j \in N.$$

In that case, the formula for z given by the second row of the dictionary makes it clear that one can minimize z over all choices of $\mathbf{x}_N \geq 0$ by letting $\mathbf{x}_N = 0$. Finally, for $\mathbf{x} = (\mathbf{x}_N, \mathbf{x}_B)$ given by $\mathbf{x}_N = 0$ and $\mathbf{x}_B = \mathbf{A}_B^{-1}\mathbf{b}$ to be a genuine optimal solution to problem (4.6), we only need to check the feasibility condition $\mathbf{x}_B \geq 0$.

We will call (4.10) the *optimality criterion* for the linear program (4.6). For our theoretical needs, this criterion is the key insight behind the simplex method. Naturally, for this criterion to be useful algorithmically, more work is needed. For example, if one of the nonbasic variables x_i has a negative coefficient in the second equation of (4.9), we need to rewrite (4.9) with x_i as a basic variable. Since the basis B has exactly m elements, we therefore need to move some variable x_j from B to N to make room for x_i. The rules for such changing of the basis are among the fundamentals of linear programming, but they need not concern us. The only additional facts we need concerning linear programming are those that support the probabilistic use of the stopping rule (4.10).

The first result is a classic existence theorem that tells us there is an optimal basic solution whenever problem (4.6) is feasible and bounded, i.e., in all the cases that matter, there is an optimal solution that satisfies (4.10). The second result we need tells us that by a modest perturbation of the cost vector, we can restrict our attention to problems for which the corresponding optimal basic solution is unique.

THEOREM 4.2.1. *If a linear program in the form (4.6) has an optimal solution, then it has an optimal basic solution, i.e., a solution of the form $\mathbf{x} = (\mathbf{x}_N, \mathbf{x}_B)$, where B is a basis and $\mathbf{x}_N = 0$.*

THEOREM 4.2.2. *If a linear program in the form (4.6) has an optimal solution, then for any $\delta > 0$ there is an ϵ with $|\epsilon| \leq \delta$ such that the corresponding problem with cost vector $\mathbf{c}' = \mathbf{c} + \epsilon$ has a unique optimal basic solution.*

The way that we will use the last "perturbation for uniqueness" theorem diverges a bit from the way that optimizers usually frame the perturbation result. More precisely, we will use the fact that for all $\{\epsilon : |\epsilon| < \delta\}$ except a set of Lebesgue measure zero, the problem with cost vector $\mathbf{c}' = \mathbf{c} + \epsilon$ has a unique optimal basic solution. This fact does not follow from Theorem 4.2.2, but it does follow from the usual proof of the theorem. We will leave this fact for individual investigation.

4.3. Dyer–Frieze–McDiarmid inequality.

The main objective of this section is to provide a bound on the expectation of the value of the solution $z^* = z^*(c_1, c_2, \ldots, c_n)$ of the linear-programming problem

$$\text{minimize} \quad z = \sum_{j=1}^{n} c_j x_j$$

(4.11)
$$\text{subject to} \quad \sum_{j=1}^{n} a_{ij} x_j = b_i, \quad i = 1, 2, \ldots, m,$$

$$x_j \geq 0, \quad j = 1, 2, \ldots, n,$$

where the cost coefficients $c_1, c_2, \ldots, c_n$ are nonnegative random variables but where the constraint coefficients a_{ij}, $1 \leq i \leq m$, $1 \leq j \leq n$, and b_i, $1 \leq i \leq m$, are fixed constants.

The bound on Ez^* is computed in terms of a *fixed* feasible solution $\hat{x}_j$, $1 \leq j \leq n$, where by "fixed" we mean that the $\hat{x}_j$'s are determined only by consideration of (a_{ij}) and (b_j), and $(\hat{x}_j)$ is not permitted to depend on the specific realization of the random cost coefficients (c_j). The general theorem that provides this bound requires a technical restriction on the distribution of the (c_j)'s, so we first consider an important special case.

THEOREM 4.3.1. *If c_j, $1 \leq j \leq n$, are independent and uniformly distributed on $[0,1]$ and $\hat{x}_j$, $1 \leq j \leq n$, is a fixed feasible solution of (4.11), then the value z^* of the optimal solution satisfies*

(4.12) $$Ez^* \leq m \max\{\hat{x}_j : 1 \leq j \leq n\}.$$

The assumption that the costs c_j have the uniform distribution enters the proof of (4.12) only through the influence that conditioning has on the expected value of a uniformly distributed random variable. Specifically, if c is a uniformly distributed random variable, then for any $0 < h < 1$ we have

(4.13) $$E(c \mid c \geq h) = h + \frac{1}{2}(1-h) = E(c) + \frac{1}{2}h.$$

The essential feature of (4.13) is that conditioning on $\{c > h\}$ boosts the expectation of c by at least a fixed fraction of h. This feature is shared by many random variables. For example, if c is exponentially distributed, then we have the analogous relation

$$E(c \mid c \geq h) = E(c) + h.$$

Guided by these two elementary observations, we are led to the statement of our main result, the Dyer–Frieze–McDiarmid inequality.

THEOREM 4.3.2. *Suppose c_j, $1 \leq j \leq n$, are independent nonnegative random variables and β, $0 < \beta \leq 1$, is a constant. If for all $h > 0$ with $P(c_j \geq h) > 0$, we have*

(4.14) $$E(c_j \mid c_j \geq h) \geq E(c_j) + \beta h,$$

then for z^*, the value of the optimal solution to the linear-programming problem (4.11), we have

(4.15) $$\beta E(z^*) \leq \max_{S:\text{card } S = m} \sum_{j \in S} \hat{x}_j E(c_j),$$

where $\hat{x}_j$, $1 \leq j \leq n$, is any fixed feasible solution of (4.11).

Proof. First, we can assume without loss of generality that the matrix $\mathbf{A} = (a_{ij})$ of constraint coefficients of (4.11) is of full rank m. The polyhedral region for (4.11) has N feasible bases $B(r) \subset \{1, 2, \ldots, n\}$, $1 \leq r \leq N$, and by the optimality criterion of (4.10), the feasible basis $B(r)$ is optimal if and only if

(4.16) $$c_j - \mathbf{c}_{B(r)} \mathbf{A}_{B(r)}^{-1} \mathbf{a}_j \geq 0 \quad \text{for all } j \in N(r),$$

where $\mathbf{a}_j$ denotes the jth column of $\mathbf{A}$. Inequality (4.16) may look innocent, but since it gives a bound on the nonbasic costs in terms of linear combinations of the basic costs, it is remarkably powerful. In fact, once the role of (4.16) is understood, the proof of the theorem just requires honest calculation of some conditional expectations.

If we let E_r denote the event that (4.16) occurs for the feasible basis $B(r)$, then the union of sets E_r, $1 \leq r \leq N$, has probability one since there exists an optimal basic solution. Moreover, since we can assume without loss of generality that with probability one the problem (4.11) has a *unique* basic solution, we also have with probability one that only one of the E_r's occurs. In other words, the E_r's form a partition of the probability space. When we apply the optimality criterion (4.10) and the conditioning hypothesis (4.14), we find a key identity that tells us the conditional expectation of c_j for $j \notin B(r)$ given that $B(r)$ is an optimal basis:

$$E(c_j \mid E_r, \mathbf{c}_{B(r)}) = E(c_j \mid c_j > \mathbf{c}_{B(r)} \mathbf{A}_{B(r)}^{-1} \mathbf{a}_j \text{ and } \mathbf{c}_{B(r)})$$
(4.17)
$$\geq E(c_j) + \beta \mathbf{c}_{B(r)} \mathbf{A}_{B(r)}^{-1} \mathbf{a}_j.$$

To exploit this fundamental fact, we need to develop an expression that relates z^* to our feasible solution $\hat{x}_j$. To move toward such an expression, we first calculate the expected value of our feasible solution given that $B(r)$ is an optimal basis. When we multiply (4.17) by $\hat{x}_j$ and sum over j, we find for each r that

$$E\left(\sum_{j=1}^n c_j \hat{x}_j \mid E_r \text{ and } \mathbf{c}_{B(r)}\right) = \sum_{j \in B(r)} c_j \hat{x}_j + \sum_{j \in N(r)} E\left(c_j \mid E_r \text{ and } \mathbf{c}_{B(r)}\right) \hat{x}_j$$

$$\geq \sum_{j \in B(r)} c_j \hat{x}_j + \sum_{j \in N(r)} \left\{E(c_j) + \beta \mathbf{c}_{B(r)} \mathbf{A}_{B(r)}^{-1} \mathbf{a}_j\right\} \hat{x}_j.$$

Now when $B(r)$ is an optimal basis, we have from the dictionary representation (4.9) that $z^* = \mathbf{c}_{B(r)} \mathbf{A}_{B(r)}^{-1} \mathbf{b}$ is the optimal value of the objective function (since we take $\mathbf{x}_N = \mathbf{0}$). To get z^* involved with (4.17), we recall that feasibility of the $\hat{x}_j$ means that

(4.18)
$$\sum_{j=1}^n \mathbf{a}_j \hat{x}_j = \mathbf{b},$$

so (4.17) gives us

$$E\left(\sum_{j=1}^n c_j \hat{x} \mid E_r, \mathbf{c}_{B(r)}\right)$$

$$\geq \sum_{j \in B(r)} c_j \hat{x}_j + \sum_{j \in N(r)} E(c_j) \hat{x}_j + \beta \mathbf{c}_{B(r)} \mathbf{A}_{B(r)}^{-1} \left(\mathbf{b} - \sum_{j \in B(r)} \mathbf{a}_j \hat{x}_j\right)$$

$$\geq \sum_{j \in B(r)} \left(c_j - \beta \mathbf{c}_{B(r)} \mathbf{A}_{B(r)}^{-1} \mathbf{a}_j\right) \hat{x}_j + \sum_{j \in N(r)} E(c_j) \hat{x}_j + \beta E(z^* \mid E_r, \mathbf{c}_{B(r)}).$$

The definitions of $\mathbf{a}_j$ and $\mathbf{A}_{B(r)}$ tell us that

$$c_j = \mathbf{c}_{B(r)} \mathbf{A}_{B(r)}^{-1} \mathbf{a}_j \quad \text{for } j \in B(r),$$

so the coefficient of $\hat{x}_j$ in the sum over $B(r)$ is just $(1-\beta)c_j \geq 0$. This gives us a bound that brings z^* out into the sunshine:

$$(4.19) \quad E\left(\sum_{j=1}^n c_j \hat{x}_j \mid E_r \text{ and } \mathbf{c}_{B(r)}\right) \geq \sum_{j \in N(r)} E(c_j)\hat{x}_j + \beta E(z^* \mid E_r \text{ and } \mathbf{c}_{B(r)}).$$

The rest of the proof is automatic. We first take expectations over the values of $c_{B(r)}$ conditionally on $B(r)$ being a feasible basis. We then write $p_r = P(E_r)$ and apply the law of total probability to find

$$\sum_r p_r E\left(\sum_{j=1}^n c_j \hat{x}_j \mid E_r\right) \geq \sum_r p_r \sum_{j \in N(r)} E(c_j)\hat{x}_j + \beta \sum_r p_r E(z^* \mid E_r).$$

In other words,

$$\sum_{j=1}^n E(c_j)\hat{x}_j \geq \sum_r p_r \sum_{j \in N(r)} E(c_j)\hat{x}_j + \beta E(z^*).$$

Finally, we have

$$\beta E z^* \leq \sum_{j=1}^n E(c_j)\hat{x}_j - \sum_r p_r \sum_{j \in N(r)} E(c_j)\hat{x}_j$$

$$= \sum_r p_r \left\{\sum_{j=1}^n E(c_j)\hat{x}_j - \sum_{j \in N(r)} E(c_j)\hat{x}_j)\hat{x}_j\right\}$$

$$= \sum_r p_r \sum_{j \in B(r)} E(c_j)\hat{x}_j \leq \max_{S:\text{card } S=m} \sum_{j \in S} E(c_j)\hat{x}_j$$

since $\operatorname{card} B(r) = m$ for all r.

Comment on a common practice.

In many "practical" contexts, the costs that figure into a linear program are actually random variables, and a common practice for dealing with such variables is to replace them with their corresponding expected values. The information that we have gathered on the assignment problem and the Dyer–Frieze–McDiarmid inequality help us see what a foolhardy practice this can be. For example, if one replaces the uniformly distributed variables c_{ij} with their expected values $Ec_{ij} = \frac{1}{2}$, then every assignment has cost $n/2$, while we know that the expected value of the optimal assignment is not larger than 2. One would be hard pressed to find a more dramatic failure of the "replace by the expected values" heuristic.

4.4. Dealing with integral constraints.

The linear program described by (4.5) seemed to differ from the assignment problem (4.1) because the x_{ij}'s were permitted to take on values other than 0 or 1. In this section, we will prove that the assignment problem is exactly equivalent to (4.1). The proof provides an engaging illustration of one of the several interesting ways that linear algebra can speak to problems of combinatorial optimization.

First, we look at the assignment problem in more graph-theoretic terms. Recall that a bipartite graph $G = (V, E)$ is one whose vertex set $V(G)$ has a partition into V_1 and V_2 such that each edge of E has one vertex in V_1 and one vertex in V_2. A compact way of writing the assignment problem is in terms of assignments of weights x_e to the edges e of a bipartite graph. Specifically, we write

$$\begin{aligned} \text{minimize} \quad & \sum_e x_e c_e \\ \text{subject to} \quad & \sum_{e: i \in e} x_e = 1 \quad \text{for all } i \in V(G) \\ \text{and} \quad & x_e \geq 0 \quad \text{with } e \in E(G). \end{aligned}$$
(4.20)

To summarize the constraints of (4.20) in a more compact form, we use the vertex-edge incidence matrix $\mathbf{A} = (a_{ve})$ defined by

$$a_{ve} = 1 \quad \text{if } v \in e \quad \text{and} \quad a_{ve} = 0 \quad \text{if } v \notin e.$$

The matrix $\mathbf{A}$ is thus a 0–1 matrix with one column for each edge and one row for each vertex of the complete bipartite graph. With $\mathbf{x} = (x_e)^T$, $\mathbf{c} = (c_e)$, and $\mathbf{1} = (1, 1, \ldots, 1)^T$, the linear-programming version of the assignment problem (4.5) can now be written as

$$\begin{aligned} \text{minimize} \quad & \mathbf{cx} \\ \text{subject to} \quad & \mathbf{Ax} = \mathbf{1} \\ \text{and} \quad & \mathbf{x} \geq 0. \end{aligned}$$

THEOREM 4.4.1. *If $\mathbf{A}$ is the vertex-edge incidence matrix of a bipartite graph G, then every square submatrix of $\mathbf{A}$ has determinant 0, 1, or -1.*

Proof. We suppose that $\mathbf{S}$ is a $k \times k$ submatrix of A and we will show that $\det \mathbf{S} \in \{0, -1, 1\}$ by using induction on k. Since $\mathbf{A}$ is a 0–1 matrix, so is $\mathbf{S}$, and the case $k = 1$ is immediate. Next, consider the possible column expansions of $\det \mathbf{S}$. Since each edge meets two vertices, each column of $\mathbf{S}$ has at most two 1's. If some column has no 1's, then $\det \mathbf{S} = 0$, and if some column has just one, we can expand about that column and proceed by induction.

Thus we can assume each column of $\mathbf{S}$ has exactly two 1's. If $V = (V_1, V_2)$ is the bipartition of the vertices of G, then the sum of the V_1-class rows of $\mathbf{S}$ is equal to $(1, 1, \ldots, 1)$ and similarly for the V_2-class rows. Thus we have a linear dependency between the rows. Consequently, the rank of $\mathbf{S}$ is less than k, and $\det \mathbf{S} = 0$.

Our interest in this purely combinatorial result is driven by its corollary, which gives us final reassurance that the optimal solution to our linear programming version of the assignment problem is indeed equal to the optimal solution to the graph-theoretical version.

COROLLARY 4.4.1. *For the vertex-edge incidence matrix* **A** *of a bipartite graph, the vertices of the polytope defined by*

$$\mathbf{Ax} \leq 1,$$
$$\mathbf{x} \geq 0$$

consists only of 0–1 vectors. Moreover, these vectors are incidence vectors of matchings.

Proof. We know that the assignment problem is bounded and feasible, so there is an optimal basic solution given by $\mathbf{x} = (\mathbf{x}_N, \mathbf{x}_B)$, where $\mathbf{x}_N = 0$ and $\mathbf{x}_B = \mathbf{A}_B^{-1}\mathbf{b}$. By the preceding theorem, all of the minors of $\mathbf{A}_B$ have determinant $0, -1$, or 1, so by the fact that $\mathbf{A}_B$ is full-rank, we see that these determinants are all ± 1. By Cramer's rule, we see that $\mathbf{A}_B^{-1}$ is an integer matrix, and since $\mathbf{b} = 1$, we see that $\mathbf{x}_B$ is an integer vector. By the constraints $0 \leq x_{ij} \leq 1$, we see that $\mathbf{x}_B$ consists only of 0's and 1's as required.

4.5. Distributional bounds.

The technique used to bound the expectation of the optimal value z^* of the objective function of the linear program with random costs can also be used to obtain information on the distribution of z^*. To get such information, we naturally need to assume more about the $\{c_j\}$'s than we did in the proof of the Dyer–Frieze–McDiarmid inequality. We will again rely on the optimality criterion of the simplex method, and thus we are led to study **c** conditioned on the event $\{c_j \geq t\}$.

Such conditioning is particularly nice when the $\{c_j\}$'s have the exponential distribution $P(c_j \geq t) = e^{-\lambda_j t}$ since we the have the lack-of-memory property:

$$P(c_j > t + s \mid c_j > t) = P(c_j > s);$$

or, more appropriately for our application, we have for all $h \geq 0$ that

(4.21) $$E(\phi(c_j) \mid c_j \geq h) = E(\phi(c_j + h))$$

provided, of course, that the expectations exist.

Since (4.21) makes our conditioning arguments as easy as they can possibly be, we will first develop distributional results for the exponential, and then we will see what these bounds tell us about other important cases. We should also note that the moment-generating function $M_j(t)$ of c_j is given by

$$Ee^{tc_j} = \frac{\lambda_j}{\lambda_j - t} \quad \text{for all } t \in [0, \lambda_j),$$

and for all $0 \leq s < \infty$, we further have

$$E(e^{tc_j} \mid c_j > s) = \frac{e^{ts}\lambda_j}{\lambda_j - t} = e^{ts}M_j(t).$$

THEOREM 4.5.1. *If the random variables c_j are independent and exponentially distributed with parameter λ_j for $1 \leq j \leq n$, then the moment-generating function $M(t)$ of the optimal solution z^* to the linear-programming problem* (4.11) *satisfies*

$$M(t) \leq \max_{S, |S|=m} \prod_{j \in S} \left\{ \frac{\lambda_j}{\lambda_j - \hat{x}_j t} \right\},$$

where $\{\hat{x}_j\}$ is a fixed feasible solution of (4.11) *and $0 \leq t < \min\{\lambda_j/\hat{x}_j, 1 \leq j \leq n\}$.*

Proof. We begin by examining the conditional moment-generating function of the cost associated with our fixed feasible solution $\{\hat{x}_j\}$, and, as before, we take E_r to be the event that $B(r) \subset [1, n]$ is an optimal basis. We first note that

$$E \left\{ \exp\left(t \sum_{j=1}^n c_j \hat{x}_j \right) \mid E_r \text{ and } \mathbf{c}_{B(r)} \right\}$$

$$= \exp\left\{ t \sum_{j \in B(r)} c_j \hat{x}_j \right\} \prod_{j \in N(r)} E\left\{ \exp(tc_j \hat{x}_j) \mid c_j \geq \mathbf{c}_{B(r)} \mathbf{A}_{B(r)}^{-1} \mathbf{a}_j \text{ and } \mathbf{c}_{B(r)} \right\}.$$

We would like to invoke the lack-of-memory property (4.21), but a little care is needed since $\mathbf{c}_{B(r)} \mathbf{A}_{B(r)}^{-1} \mathbf{a}_j$ could be negative. Using $(w)_+$ to denote the positive part of w, the lack-of-memory property tells us that the last product equals

$$\exp\left(\sum_{j \in B(r)} tc_j \hat{x}_j + \sum_{j \in N(r)} t\hat{x}_j \left(\mathbf{c}_{B(r)} \mathbf{A}_{B(r)}^{-1} \mathbf{a}_j \right)_+ \right) \prod_{j \in N(r)} M_j(\hat{x}_j t)$$

$$\geq \exp\left(\sum_{j \in B(r)} tc_j \hat{x}_j + \sum_{j \in N(r)} t\hat{x}_j \mathbf{c}_{B(r)} \mathbf{A}_{B(r)}^{-1} \mathbf{a}_j \right) \prod_{j \in N(r)} M_j(\hat{x}_j t)$$

$$= \exp\left(t\mathbf{c}_{B(r)} \mathbf{A}_{B(r)}^{-1} \mathbf{b} \right) \prod_{j \in N(r)} M_j(\hat{x}_j t),$$

where in the last expression we used the fact that the feasibility of $\hat{x}_j$ tells us precisely that $\sum \mathbf{a}_j \hat{x}_j = \mathbf{b}$.

We can then take expectations over $\mathbf{c}_{B(r)}$ and recall that $z^* = \mathbf{c}_{B(r)} \mathbf{A}_{B(r)}^{-1} \mathbf{c}$ to find

$$(4.22) \quad E\left\{ \exp\left(t \sum_{j=1}^n c_j \hat{x}_j \right) \mid E_r \right\} \geq E\left\{ \exp(tz^*) \prod_{j \in N(r)} M_j(\hat{x}_j t) \mid E_r \right\}.$$

Now with probability one, we have the trivial inequality

$$\exp(tz^*) \prod_{j \in N(r)} M_j(\hat{x}_j t) \geq \exp(tz^*) \min_T \left\{ \prod_{j \in T} M_j(\hat{x}_j t) : \text{card } T = n - m \right\},$$
(4.23)

and since the last product in (4.23) does not depend on E_r, we can use this lower bound in (4.22) and take the expectation over the conditioning to find

$$\prod_{j=1}^{n} M_j(\hat{x}_j t) \geq E(\exp(tz^*)) \min_T \left\{ \prod_{j \in T} M_j(\hat{x}_j t) : \text{card } T = n - m \right\}.$$

On rearranging the product terms, we find the desired inequality

$$E \exp(tz^*) \leq \max_{S : \text{card } S = m} \prod_{j \in S} M_j(\hat{x}_j t).$$

A bound on a moment-generating function can be exploited in a many ways, but for the moment we note only the simplest consequence.

COROLLARY 4.5.1. *Suppose c_j, $1 \leq j \leq n$, are independent random variables with a distribution $F(t)$ satisfying*

$$1 - F(t) \leq e^{-\lambda t}$$

for all $t \geq 0$. If $(\hat{x}_1, \hat{x}_2, \ldots, \hat{x}_n)$ is a fixed feasible solution of (4.11), then for $x = \max_{1 \leq j \leq n} \hat{x}_j$ and any $\delta > 0$, we have

$$P(z^* \geq (1+\delta) m x / \lambda) \leq \exp(-m(\delta - \log(1+\delta))).$$

Here we should note that the last inequality applies with $\lambda = 1$ for independent and uniformly distributed c_j since for all $x \geq 0$, we have $P(c_j \geq x) = 1 - x \leq e^{-x}$.

4.6. Back to the future.

We used the Dyer–Frieze–McDiarmid inequality to obtain the bound $EA_n \leq 2$ due to Karp (1983, 1987), who first showed how to prove useful bounds on linear-programming solutions by conditioning on an optimal basis. The conditioning method is sure to have many future uses, so we would do well to go back to the original application to see what might be learned. Karp's approach is quite close to that of Dyer, Frieze, and McDiarmid, but it differs in some important details, at least so far as Karp made use of the dual and of the explicit structure of the feasible bases.

One fact that we have not used before is that the assignment problem is a special case of the *transportation problem*:

$$\begin{aligned}
\text{minimize} \quad & \sum_{i=1}^{m} \sum_{j=1}^{n} c_{ij} x_{ij} \\
\text{subject to} \quad & x_{ij} \geq 0, \\
& \sum_{j=1}^{n} x_{ij} = a_i \quad \text{for } i = 1, 2, \ldots, m, \\
& \sum_{i=1}^{n} x_{ij} = b_j \quad \text{for } j = 1, 2, \ldots, n.
\end{aligned}$$

92   CHAPTER 4

Moreover, the transportation problem has been studied extensively, and much is known about its structure. In particular, we know that the dual for the transportation problem can be written as

$$\text{maximize} \quad \sum_{i=1}^{m} a_i u_i + \sum_{j=1}^{n} b_j v_j$$
$$\text{subject to} \quad c_{ij} - u_i - v_j \geq 0 \quad \text{for all } i = 1, 2, \ldots, m \text{ and } j = 1, 2, \ldots, n.$$

We should note here that for either the primal or dual problem to make sense, we need to have the balance condition

$$\sum_{i=1}^{m} a_i = \sum_{j=1}^{n} b_j.$$

To get to the assignment problem from the transportation problem, one only needs to impose the additional restrictions that $n = m$ and $a_i = b_j = 1$ for all $1 \leq i, j \leq n$.

One of the important features of the transportation problem is that the basic solutions to both the primal and the dual problem can be expressed in terms of spanning trees. To make this explicit for the dual problem, we first introduce G for the complete bipartite graph with bipartition $V_1 = \{w_1, w_2, \ldots, w_n\}$ and $V_2 = \{w'_1, w'_2, \ldots, w'_n\}$. With probability one, a realization of c_{ij} provides an instance of the assignment problem for which there is a unique spanning tree T of G such that the optimality conditions for the dual problem can be written in terms of T:

(4.24) $\qquad c_{ij} - u_i - v_j \geq 0 \quad \text{for all } 1 \leq i, j \leq n$

and

(4.25) $\qquad c_{ij} - u_i - v_j = 0 \quad \text{for all } i, j \text{ such that } (w_i, w'_j) \in T.$

These relations provide the foundation for our conditioning argument.

We let $A(T, u, v)$ denote the event that a spanning graph T of G satisfies (4.24) and (4.25). Karp's key observation is that, conditional on the event $A(T, u, v)$, the distribution of c_{ij} for $(w_i, w'_j) \notin T$ is uniformly distributed on the set $[\max\{0, u_i + v_j\}, 1]$, but all that we use from this fact is that we have the conditional expectation

(4.26) $\qquad (w_i, w'_j) \notin T \Rightarrow E(c_{ij} | A(T, u, v)) = \frac{1}{2} + \frac{1}{2} \max\{0, u_i + v_j\}.$

The rest of the argument is just a pleasant computation. Specifically, by (4.26), we have

$$S(T, u, v) \equiv E\left\{ \sum_{i=1}^{n} \sum_{j=1}^{n} c_{ij} | A(T, u, v) \right\}$$
$$= \sum_{(w_i, w'_j) \in T} \{u_i + v_j\} + \sum_{(w_i, w'_j) \notin T} \frac{1}{2} + \frac{1}{2} \max(0, u_i + v_j).$$

By the crude bound $\max(0, u_i + v_j) \geq u_i + v_j$, we find

$$S(T, u, v) \geq \sum_{(w_i, w'_j) \in T} \{u_i + v_j\} + \sum_{(w_i, w'_j) \notin T} \frac{1}{2} + \frac{1}{2}(0, u_i + v_j)$$

$$= \frac{1}{2} \sum_{i=1}^{n} \sum_{j=1}^{n} \{u_i + v_j\} + \frac{1}{2} \{n^2 - 2n + 1\} = \frac{1}{2} n A_n + \frac{1}{2} \{n^2 - 2n + 1\},$$

where in the last step we used the fact that for the dual problem we have

$$A_n = \sum_{i=1}^{n} u_i + \sum_{j=1}^{n} v_j.$$

Now if we use $E c_{ij} = \frac{1}{2}$ to calculate the unconditional expectation and the law of total probability to compute the expectation of the lower bound, we find

$$ES(T, u, v) = \frac{1}{2} n^2 \geq \frac{1}{2} n E A_n + \frac{1}{2} \{n^2 - 2n + 1\},$$

from which we find Karp's inequality:

$$EA_n \leq (2n - 1)/n < 2.$$

Here we should note some of the differences between this proof and the argument of Dyer, Frieze, and McDiarmid. First, the result is just a hair stronger— the Dyer–Frieze–McDiarmid inequality gave precisely $EA_n \leq 2$, so there are differences to be found.

4.7. Additional remarks.

The theory of random graphs and the theory of linear programming both have a huge literature, and there are many additional results that we could have covered in this chapter. The following notes touch on only those issues that directly engage the material that was covered.

1. McDiarmid (1986) has further explored the application of the conditioning method to optimization problems associated with the greedy algorithm.

2. The proof of Walkup's inequality given here does not follow Walkup (1979) in detail, though the method is essentially the same. The proof is patterned more closely after an analysis given in Karp, Rinnooy Kan, and Vohra (1995), except that some details have been added to the conditioning argument. The main theme of Karp, Rinnooy Kan, and Vohra (1995) is the demonstration of a $O(n \log n)$ heuristic which with high probability will find a perfect matching in a random two-out graph. When one applies this heuristic to the assignment problem with independent uniform costs, one finds an assignment that with high probability is not larger than $3 + O(n^{-a})$ for an $a > 0$. Avis and Devroye (1985) and Avis and Lai (1988) also provide a heuristic that with high probability will lead to assignments with bounded expected cost.

3. Frenk, van Houweninge, and Rinooy Kan (1987) investigated the assignment problem with independent cost but with more a general distribution F supported

on $(-\infty, \infty)$, or $(0, \infty)$. Under mild restrictions on F, the expected value of the optimal assignment was found to be both $O(nF^{-1}(1/n))$ and $\Theta(nF^{-1}(1/n))$. The these results were based on arguments like those used by Walkup (1979).

4. The discussion of the dictionary form for the standard linear program is based on the elegant development of the simplex method given in the text of Chvátal (1980).

5. Bertsekas (1991) gives an extensive discussion of algorithms for the assignment problem, including computer programs for the calculation of the minimum cost assignment.

6. The lower bound $\limsup E(A_n) \geq 1 + e^{-1}$ was proved by Lazarus (1979), and at the cost of considerable effort, this result was sharpened by Goemans and Kodialam (1993) to $\limsup E(A_n) \geq 1 + e^{-1} + \iota$ for a small but explicit constant $\iota > 0$.

Since one has $E(A_n) < 2$, one is naturally led to believe that $E(A_n)$ converges to a constant. This result has been proved by Aldous (1992) on the basis of a sustained argument using his *objective method*, a creative variation on the theme of weak convergence where one extracts information about finite objects like A_n by the construction of an appropriate infinite analogue. We will study the application of the objective method to a problem in the theory of minimum spanning trees in the next chapter.

One naturally conjectures that $E(A_n)$ is a monotone increasing sequence. This fact may even be easy, but so far a proof has not been found. Quite possibly, an appropriate linear-programming representation for A_n will provide the missing step. The more fascinating (but less likely) possibility is that $E(A_n)$ is not monotone.

7. Mézard and Parisi (1987) have provided an engaging but nonrigorous argument based on ideas from statistical mechanics that $E(A_n) \to \pi^2/6$. A rigorous proof or disproof of the suggested limit result would be of great interest.

CHAPTER 5

Distributional Techniques and the Objective Method

This chapter develops two distributional techniques that are well adapted for application to probability problems of combinatorial optimization. The first of these is the objective method, which offers a way of thinking about the theory of weak convergence that requires the development of infinite analogues to finite combinatorial objects. Our test case for the objective method is the resolution of a delicate problem from the theory of minimal spanning trees. The second technique is the conditioning method for obtaining d-dependence. This method can be used to provide central limit theorems for problems like the length of the nearest-neighbor matching, the length of the boundary of the Voronoi diagram, and the length of the Delaunay triangulation.

5.1. Motivation for a method.

If $S = \{x_1, x_2, \ldots, x_n\}$ is a finite set of points in $\mathbb{R}^d$, $d \geq 2$, then we recall that $T = T(S)$ is a minimal spanning tree (MST) of S if $T(S)$ is a connected graph with vertex set S such that the sum of the edge lengths of T is minimal; that is, T is a minimal spanning tree for S if T satisfies

$$\sum_{e \in T} |e| = \min_G \sum_{e \in G} |e|,$$

where $|e| = |x_i - x_j|$ is the Euclidean length of the edge $e = (x_i, x_j)$ and the minimum is over all connected graphs G with vertex set S. We have already seen some illustrations of the fact that minimal spanning trees have a fundamental role the theory of computational geometry. They are also among the most studied objects in combinatorial optimization.

One of the motivations for the first problem that we explore in this chapter came from a simulation experiment. R. Bland observed in the course of studying algorithms for the MST that when the $\{X_i\}$'s are independent with the uniform distribution on the unit square, one seems to have

$$\sum_{e \in \text{MST}} |e|^2 \to c > 0.$$

This conjecture fits naturally as part of the broader problem of minimal spanning trees with power-weighted edges; and, motivated in part by Bland's conjecture, Steele (1988) proved that if the random variables $\{X_i\}$ are independent and with a distribution μ with compact support, then for all $0 < \alpha < d$ one has

$$(5.1) \quad n^{-(d-\alpha)/d} \sum_{e \in \text{MST}} |e|^\alpha \to c(\alpha, d) \int_{\mathbb{R}^d} f(x)^{(d-\alpha)/d}\, dx \quad \text{a.s.,}$$

where f is the density of the absolutely continuous part of μ and $c(\alpha, d)$ is a constant that depends only on α and d. The method used in Steele (1988) was based on subadditivity arguments like those we have used in several of the earlier chapters, and an interesting limitation of the straightforward application of that technique is that it seems to leave the critical case $\alpha = d$ just out of its reach. The first goal of this chapter is to explain an approach to this problem that was developed in Aldous and Steele (1992) to handle the critical case. The approach is based on the "objective method," which one might regard as more of a philosophy than an explicit technique. Still, even as a philosophy, the objective method seems to offer a useful way to think about some of the harder probability problems of combinatorial optimization.

At the heart of the objective method is the notion that some questions about an increasing sequence of finite random objects can be studied by inventing an infinite random object for which one has some good analogue to the question that is of concern for the sequence of finite objects. If an appropriate infinite object can be found, then there is every hope that the limiting behavior of the finite objects can be read out of the stationary behavior of the infinite object. Naturally, this suggestion sounds vague (it is!), but the philosophy still gives one a plan for progress. There have already been several successes with this plan, and there is a richness to the method that suggests that there will be more to come.

In the case of the MST, the objective method tells us to look for an analogue of the MST for an *infinite* set of points. Next, since a finite sample of independent uniformly distributed points in $[0,1]^d$ looks locally like Poisson processes, the natural idea is to relate the MST of finite samples in $[0,1]^d$ to the MST of the unbounded Poisson process in $\mathbb{R}^d$—provided, of course, that we can produce a good candidate for such an MST.

5.2. Searching for a candidate object.

Before we engage the probability theory of our infinite MST, we need some elementary facts about the convergence of sequences of discrete sets. We first note that we usually have no loss of generality if we restrict our attention to certain *nice* sets. Specifically, for $S = \{x_i\} \subset \mathbb{R}^d$ with $d \geq 2$, we will call S *nice* if S is *locally finite* (i.e., S has only finitely many elements in any bounded subset of $\mathbb{R}^d$) and if $S = \{x_i\}$ has the property that all of the interpoint distances $\{|x_j - x_i|,\ i < j\}$ are distinct. If we are studying a finite sample from a continuous distribution or a realization of the Poisson process in $\mathbb{R}^d$, there is no loss of generality in assuming that our processes produce only nice realizations. Thus for the rest of the chapter, we will restrict attention exclusively to nice sets.

DISTRIBUTIONAL TECHNIQUES AND THE OBJECTIVE METHOD 97

Now we officially begin our hunt for candidate for the idealization of an "MST on an infinite random set in $\mathbb{R}^d$." Given a pair (x, S) with S nice and $x \in S$, we can define a sequence of trees $T_m = T_m(x, S)$ with vertices from S by a sequential process:

- To begin, we let $X_1 = x$, and then we just take our first tree T_1 to be the singleton vertex set $\{X_1\}$ and the empty edge set.

- Next, we let T_2 be the tree with vertex set consisting of X_1 together with the vertex $X_2 \in S \setminus \{X_1\}$ which is closest to X_1 in Euclidean distance, and for the edge set of T_2, we just take the single the edge connecting X_1 and X_2.

- We then proceed inductively by defining $T_m = T_m(x, S)$ to be the tree T_{m-1} together with a new edge (X_{j_m}, X_m), where $j_m \leq m - 1$ and a new point $X_m \in S \setminus \{X_1, \ldots, X_{m-1}\}$ are chosen so that the edge length $|X_m - X_{j_m}|$ is minimal over all edges that connect T_{m-1} to the complement of T_{m-1} in S.

In the case where $n = \text{card}\, S < \infty$, this procedure terminates with a tree $T_n(x, S)$ that has all of S as its vertex set. This tree is, in fact, the unique minimal spanning tree of the nice set S. We note as a consequence of this uniqueness for nice finite S that the terminal tree $T_n(x, S)$ for such a set does not depend on the choice of the starting vertex x. The case where $\text{card}\, S = \infty$ is far more interesting. Even to discuss the "terminal tree" for an infinite S, we need to make a definition, and of course it is natural to take

$$T_\infty(x, S) \equiv \bigcup_n T_n(x, S).$$

Now if $T_\infty(x, S)$ would serve honorably as our candidate for the minimal spanning tree of S, we would be well on our way. Unfortunately, this candidate is not good enough. In particular, one can easily construct a nice infinite S for which the vertex set of $T_\infty(x, S)$ does not cover all of S. Thus if we are to have a useful candidate for the minimal *spanning* tree of an infinite S, we must be willing to accept a more complex construction.

The next variation one might explore is to take the union of $T_\infty(x, S)$ over all $x \in S$. This is certainly a spanning graph, but it is not evident that this union does not contain cycles, and the occurrence of cycles would provide a bad start for a graph that is suppose to reflect the behavior of a spanning *tree*. Still, with just one more turn, we will have pressed out a successful choice.

A spanning forest.

Although neither the individual trees $T_\infty(x, S)$ nor their union over all x's seems to serve well as a candidate for an infinite MST object, they point to a graph that does serve well. For the vertex set S, the graph $G = G(S)$ of interest will be defined by taking the edge set as that subset E of the complete set of edges on S that are chosen by the following rule:

(5.2) $\quad e = (x_1, x_2) \in E \quad \Longleftrightarrow \quad e \in T_\infty(x_1, S) \quad \text{or} \quad e \in T_\infty(x_2, S).$

Key facts about the current candidate.

The new candidate is certainly a spanning graph. To check that G contains no cycles, we can argue by contradiction. Suppose that $y_1 \to y_2 \to \cdots \to y_n \to y_1$ is a cycle in G, and suppose further that $e = (y_1, y_n)$ is the longest edge in the cycle. We first note that $e \notin T_\infty(y_1, S)$ since there is a path from y_1 to y_n that uses edges that are all shorter than $|y_1 - y_n|$. The same story applies to show that $e \notin T(y_n, S)$, so by the definition of G, we have $e \notin G$. Thus we see that G has no cycles, so by definition G is a forest—a union of disjoint trees. Naturally, G would be a tree if G were connected, and to show at least that we have a high degree of connectivity, we next check that each of the connected components of G is infinite. For this purpose, we will benefit from a simple lemma.

LEMMA 5.2.1. *If (y_1, y_2) is an edge of $T_\infty(x, S)$ for some $x \in S$, then (y_1, y_2) is also an edge of either $T_\infty(y_1, S)$ or $T_\infty(y_2, S)$.*

Proof. Consider the graph H with vertex set S and edge set consisting of all $e = (x, y)$ with $|x - y| < |y_1 - y_2|$. To prove the contrapositive of the statement of the lemma, we first suppose that $(y_1, y_2) \notin T_\infty(y_1, S)$ and that $(y_1, y_2) \notin T_\infty(y_2, S)$. The first statement implies that y_1 is an infinite component of H, and the second statement does the same job for y_2.

We can suppose that both y_1 and y_2 are in $T_\infty(x, S)$, or else we certainly cannot have (y_1, y_2) as an edge of $T_\infty(x, S)$. Let O be the ordering the points of $T_\infty(x, S)$ according to their addition to its construction. Since y_1 is in an infinite component of H, all of the points that follow y_1 in the order O were added with an edge of length less than $|y_1 - y_2|$. Since the same is also true of y_2, we see that the edge (y_1, y_2) can never be added.

The real punchline that we can draw from the this lemma is that we now see, as promised, that each connected component C of our forest G must be infinite. To see this, we just note that G has no isolated points and that if (y_1, y_2) is an edge of C, then the lemma tells us that at least one of the infinite sets $T_\infty(y_1, S)$ and $T_\infty(y_2, S)$ is contained in C.

We have thus shown a strong family resemblance between G and what we would want in a minimal spanning tree for an infinite set S. If we could show for the sets S of interest to us that the graph is connected, then G would indeed be a perfect candidate. Even though we will not show here that our construction leads to a connected G, we will still be able to extract useful information from the graph.

The final candidate and two results.

The lead role in the rest of this chapter is now given over to the a random tree T that we build out of the Poisson process and our forest $\mathcal{G}$. Specifically, we take a Poisson point process $\mathcal{N}$ of rate 1 in $\mathbb{R}^d$, $d \geq 2$, and we further set $\mathcal{N}^0 = \mathcal{N} \cup \{0\}$. With probability one, $\mathcal{N}^0$ is a nice subset of $\mathbb{R}^d$, and as we remarked earlier, we will without loss of generality restrict attention to realizations of the Poisson process that are *always* nice sets. Next, we let $\mathcal{G}$ be the forest given by our construction when it is applied to $\mathcal{N}^0$, and as the idealized minimal spanning tree T, we take the component of $\mathcal{G}$ that contains the vertex 0. The extent to which T is a wise choice depends only on whether T teaches us what we want to know.

Almost anything that one might learn about T is of interest since we have at our disposal the tools that let us map properties of T into limit theorems of the kind that have been our central concern for the most previous chapters. In the two theorems that follow, we do see that T serves usefully as an infinite object whose charge is to inform us about the limit behavior of sequences of finite objects. The first of these theorems resolves the conjecture of R. Bland that was mentioned in the introduction.

THEOREM 5.2.1. *If the random variables $\{X_i\}$ are independent with the uniform distribution on $[0,1]^d$ and $S^n = \{X_i : 1 \leq i \leq n\}$, then we have for all $p \geq 1$ that*

$$\sum_{e \in \text{MST}(S^n)} |e|^d \to l_d \quad \text{in } L^p \quad \text{as } n \to \infty,$$

where

(5.3) $$l_d \equiv \frac{1}{2} \sum_i E L_i^d < \infty$$

and the $\{L_i\}$'s are the lengths of the edges of T that are incident to the point 0.

This theorem not only tells us that we have the conjectured convergence, but we also learn something about the limit constant. We are not so lucky as to obtain a numerical value, but we do find what might be just as valuable—an interpretation of the constant as a fixed quantity in a model where no limits need be taken. The second limit result that we extract from our illustration of the objective method tells us about the degree sequence of the MST of a random sample.

THEOREM 5.2.2. *If D is the degree of vertex 0 in T, then*

(5.4) $$ED = 2,$$

and if the random variables $\{X_i\}$ are independent with the uniform distribution on $[0,1]^d$ and $S^n = \{X_i : 1 \leq i \leq n\}$, then we also have

(5.5) $$\frac{1}{n} E \text{card}\{x \in S^n : \deg(x, \text{MST}(S^n)) = i\} \to P(D = i).$$

The fact that $ED = 2$ may seem surprising at first—we find so few exact constants—but the identity just reflects an appropriate $n = \infty$ interpretation of the fact that in any tree with n vertices, we have $n-1$ edges, so the average degree is exactly $2(n-1)/n$. The existence of the limit in (5.5) was already proved in Steele, Shepp, and Eddy (1987). In fact, that article proved that there is convergence with probability one for all $\{X_i\}$'s with an absolutely continuous distribution on $\mathbb{R}^d$. Still, (5.5) contains new information since it provides an interpretation of the limit constant in terms of the random variable D. The proofs of Theorems 5.2.1 and 5.2.2 will follow from more detailed results that will be developed over the next several sections.

Finally, we should note that since T does serve well as a surrogate for the MST of the Poisson process, we almost have to suspect that the graph $\mathcal{G}$, which we pessimistically called a forest, must really be a single *tree*. The natural conjecture that $\mathcal{T} = \mathcal{G}$ was put forth in Aldous and Steele (1992) and was recently proved by Alexander (1995a). We will discuss Alexander's results in a later section that deals with the central limit theory for the MST and related problems.

5.3. Topology for nice sets.

We take the convergence $S^n \to S$ of a sequence of locally finite sets S^n to a locally finite set S to mean that one can label the points of S as $x_1, x_2, \ldots$ and label the points of S^n as $x_{n,1}, x_{n,2}, \ldots$ in such a way that as $n \to \infty$, first, we have

(5.6) $$x_{n,i} \to x_i \quad \text{for all } 0 \leq i < \infty$$

and second, for each real $L > 0$ such that S has no point on the boundary of $C_L \equiv [-L, L]^d$, we have

(5.7) $$\text{card } S^n \cap C_L \to \text{card } S \cap C_L.$$

In addition to this notion of convergence of locally finite sets, we need a notion of convergence for graphs that are defined on locally finite sets. Some additional care is required in formulating an appropriate condition, but one does not have to look too hard to find a definition that will serve our needs. If H_n and H are graphs with respective vertex sets S^n and S, we define the *convergence of graphs* $H_n \to H$ to mean that we have the three following properties:

1. *Set convergence.* $S^n \to S$.
2. *H_n edges are H edges in the limit.* For each L with $S \cap \partial C_L = \emptyset$, we have for all sufficiently large n that if $(x_{n,i}, x_{n,j})$ is an edge of H_n with $x_{n,i} \in C_L$, then (x_i, x_j) is an edge of H.
3. *H edges are H_n edges in the limit.* For each L with $S \cap \partial C_L = \emptyset$, we have for all sufficiently large n that if (x_i, x_j) is an edge of H with $x_i \in C_L$, then $(x_{n,i}, x_{n,j})$ is an edge of H_n.

A most useful consequence of this definition is that $H_n \to H$ and $x_{n,i} \to x_i$ imply that $\deg(x_{n,i}, H_n)$ converges to $\deg(x_i, H)$. Naturally, our definition of graph convergence would not be a very good if we did not have this basic property.

We now recall the trees $T_k(x, S)$ that were defined earlier in the chapter, and we introduce finite graphs $G_k(S)$ in parallel to our earlier definition of the forest $\mathcal{G}$; that is, we take (x_i, x_j) as an edge in $G_k(S)$ if and only if (x_i, x_j) is an edge in either $T_k(x_i, S)$ or $T_k(x_j, S)$. For nice sets S^n and S with $\text{card } S = \infty$, we then see that for all fixed k we have the implication

$$S^n \to S \quad \text{and} \quad x_n \in S^n, \ x \in S \quad \text{with } x_n \to x \Rightarrow T_k(x_n, S^n) \to T_k(x, S).$$

We therefore see that for each fixed k, $S^n \to S$ implies that $G_k(S^n)$ and $G_k(S)$ satisfy the three conditions of graph convergence, so $G_k(S^n) \to G_k(S)$ as $n \to \infty$. We can express this last observation by saying that $G_k(\cdot)$ is a continuous function.

The forest $G(S)$ that we defined earlier in the construction of our candidate $\mathcal{T}$ for a spanning "tree" for S can also be written as $\cup_k G_k(S)$. The map $G(\cdot)$ is more subtle than the maps $G_k(\cdot)$, but some of the behavior of $G_k(\cdot)$ carries over to $G(\cdot)$. In particular, from $G_k(S^n) \to G_k(S)$ and the representation $\cup_k G_k(S)$, we find a type of semicontinuity of for the map $G(\cdot)$ that can be spelled out as follows: if we have $S^n \to S$ and (x_i, x_j) is an edge of $G(S)$ with $x_i \in C_L$, then $(x_{n,i}, x_{n,j})$ is an edge of $G(S^n)$ for all sufficiently large n.

To check that this semicontinuity of $G(\cdot)$ can not be pushed to full continuity, we only need to study a simple example. We write $\alpha_n = \sum_{i=1}^{n} 1/i$ and consider the sequence of one-dimensional sets

$$(5.8) \qquad S^n = \left\{-\sqrt{2}\alpha_n, \ldots, -\sqrt{2}\alpha_1, \alpha_1, \alpha_2, \ldots, \alpha_n\right\}.$$

By the irrationality of $\sqrt{2}$, we see that S^n is a nice set; and clearly the edge $(-\sqrt{2}, 1)$ is $G(S^n)$ for all n, yet it is not in $G(S)$. Thus we see that we do not have $G(S^n) \to G(S)$ even though $S^n \to S$.

The next lemma summarizes the preceding discussion of the convergence of nice sets. The first statement of the lemma just asserts the semicontinuity of the mapping G. The second statement of the lemma then tells us that if we supplement semicontinuity with the degree bound (5.10), then we get genuine continuity of G at S.

LEMMA 5.3.1 (semicontinuity of $G(\cdot)$ and graph convergence). *If we have $S^n \to S$, where S^n and S are nice point sets and S is infinite, then for each L such that S has no point on the boundary of C_L, we have*

$$(5.9) \qquad \liminf_n \sum_{x_{n,i} \in C_L} \deg(x_{n,i}, G(S^n)) \geq \sum_{x_i \in C_L} \deg(x_i, G(S)).$$

Moreover, if we have for each L such that $S \cap \partial C_L = \emptyset$ that

$$(5.10) \qquad \limsup_n \sum_{x_{n,i} \in C_L} \deg(x_{n,i}, G(S^n)) \leq \sum_{x_i \in C_L} \deg(x_i, G(S)),$$

then

$$G(S^n) \to G(S).$$

The lemma may look technical, but it is easy to use. For example, suppose we have a probability model under which we have $S^n \to S$ with probability one. In order to show that the graph convergence $G(S^n) \to G(S)$ also takes place with probability one, we just need to show that (5.10) holds with probability one. Thus Lemma 5.3.1 reduces the abstract problem of proving graph convergence to the more concrete task of proving a numerical inequality like (5.10). We next record a simple real-variable lemma that helps the two halves of Lemma 5.3.1 work together.

LEMMA 5.3.2. *If the random variables $Y_n \geq 0$ satisfy*

$$\liminf_{n \to \infty} Y_n \leq Y \quad a.s. \quad \text{and} \quad \lim_{n \to \infty} EY_n = EY,$$

then Y_n converges to Y in L^1.

Proof. We write $Y_n - Y = (Y_n - Y)_+ - (Y_n - Y)_-$ and note that we have the dominating bound $0 \leq (Y_n - Y)_- \leq Y$. The first hypothesis tells us that $(Y_n - Y)_- \to 0$ with probability one, so by the dominated-convergence theorem, we have $E(Y_n - Y)_- \to 0$. By the second hypothesis, we then find $E(Y_n - Y)_+ \to 0$, netting out to give the L^1 convergence, $E|Y_n - Y| \to 0$.

The point of the lemma is that with (5.9) guaranteed, all that one needs to make progress toward the verification of (5.10) is information about expectations. Since sums and expectations are the probabilist's stock and trade, we will find that this reduction provides us with an effective exit from otherwise tedious consideration of set convergence.

5.4. Information on the infinite tree.

In any application of the objective method, we must have relevant information about our infinite object before we can hope to make any useful inferences about our finite objects. Thus we will need to study the number and lengths of the edges that are incident to a typical point of our forest $\mathcal{G}$. We recall that $\mathcal{N}^x$ is a *Palm process* for the point process $\mathcal{N}$ if for all x the distribution of $\mathcal{N}^x$ is the same as the conditional distribution of $\mathcal{N}$ given that there exists a point of $\mathcal{N}$ at x.

LEMMA 5.4.1. *If $\mathcal{N}$ is a nice stationary ergodic point process and $\mathcal{N}^x$ is the associated Palm process, then the distribution of $D = \deg(x, G(\mathcal{N}^x))$ does not depend on x and $ED = 2$.*

Proof. The fact that distribution of $D = \deg(x, G(\mathcal{N}^x))$ does not depend on x is immediate from stationarity, so we just focus on the proof of the formula $ED = 2$. As before, we set $C_L = [-L, L]^d$, and then we claim that for all realizations of the process, we have the identity

$$(5.11) \qquad \sum_{X \in C_L \cap \mathcal{N}} (\deg(X, \mathcal{G}) - 2) = B_L - 2F_L,$$

where B_L is the number of edges of $G(\mathcal{N})$ that cross the boundary of C_L and F_L is the number of distinct trees of the subforest $\mathcal{F}$ of $G(\mathcal{N})$ consisting of all of the edges of $G(\mathcal{N})$ that have at least one vertex inside C_L. To prove (5.11), we recall that for any graph the sum of the degrees is equal to twice the number of edges, and in the case of $\mathcal{F}$, this says

$$\sum_{X \in C_L \cap \mathcal{N}} \deg(X, \mathcal{G}) + B_L = 2 \left\{ \sum_{X \in C_L \cap \mathcal{N}} \deg(X, \mathcal{G}) + B_L - F_L \right\},$$

where on the right-hand side we used the fact that in a forest the number of edges is equal to the number of vertices minus the number of connected components. Just by rearranging the last identity, we find (5.11).

Next, by the definition of stationarity and the intensity ρ, we have

$$E \operatorname{card} C_L \cap \mathcal{N} = \rho(2L)^d,$$

so taking expectations, we have
$$E \sum_{X \in \mathcal{N} \cap C_L} \deg(X, \mathcal{G}) = \rho(2L)^d ED.$$

Since $F_L \leq B_L$, our counting identity (5.11) will therefore imply the lemma if we can just show that $EB_L = o(L^d)$. To see why this is "obvious," we note that any edge that crosses ∂C_L is either pretty long or has an endpoint pretty close to ∂C_L—two phenomena that we can make $o(L^d)$. To be more formal, we let $D_r(x)$ denote the set of edges in $\mathcal{N}^x$ that are of the form (x, y) with $|x - y| \geq r$. By invariance, $ED_r(x) = ED_r(0)$ does not depend on x, so if we bound the number of the edges of B_L by the number of points in $C_L \setminus C_{L-r}$ together with the number of points of C_{L-r} that are incident to an edge of length at least r, then

(5.12) $\quad EB_L \leq \rho b_d E\operatorname{card}\left(\{C_L \setminus C_{L-r}\} \cap \mathcal{N}\right) + \rho 2^d (L - r)^d ED_r(0),$

where the factor $b_d < \infty$ denotes the maximum degree of any MST in $\mathbb{R}^d$—the finiteness of b_d being assured by the fact that by the triangle inequality an MST cannot have an angle between any two edges that is less than $60°$. One could therefore take b_d as the least number of spherical caps with solid angle less than $60°$ that suffice to cover the surface of the unit sphere in $\mathbb{R}^d$.

When we let $L \to \infty$ in (5.12), we find that for all $r > 0$,
$$\limsup_{L \to \infty} EB_L/(2L)^d \leq \rho ED_r(0);$$

and, since $ED_r(0) \to 0$ as $r \to \infty$, the last inequality tells us that $EB_L/(2L)^d \to 0$, just as required.

5.5. Dénoument.

We have almost all of the structure in place that we need to complete the proofs of Theorems 5.2.1 and 5.2.2. We recall that for point processes $\mathcal{N}_n$, there are several equivalent ways to define convergence in distribution, but will just use the definition that says

(5.13) $\quad\quad\quad\quad\quad\quad\quad\quad \mathcal{N}_n \xrightarrow{d} \mathcal{N}$

if and only if there are versions $\mathcal{N}'_n$ and $\mathcal{N}'$ of the processes $\mathcal{N}_n$ and $\mathcal{N}$ such that

(5.14) $\quad\quad\quad\quad\quad\quad\quad \mathcal{N}'_n(\omega) \to \mathcal{N}'(\omega) \quad \text{a.s.,}$

where in the last expression we understand that we have convergence in the sense of locally finite sets. Convergence in distribution of random graphs H_n and H with vertex sets given by point processes is defined in the same way, except that instead of (5.14), one requires that

$$H_n(\mathcal{N}'_n) \to H(\mathcal{N}') \quad \text{a.s.,}$$

where convergence in the last expression takes place in the sense of convergence of graphs defined on nice point sets. As before, we let $\mathcal{N}^0 = \mathcal{N} \cup \{0\}$, where $\mathcal{N}$ is the Poisson process of intensity 1; equivalently, $\mathcal{N}^0$ has the distribution of $\mathcal{N}$ conditional on the event that $0 \in \mathcal{N}$.

LEMMA 5.5.1. *Let $\mathcal{N}_n$ denote the point process consisting of n points $\{X_i : 1 \leq i \leq n\}$ which are independent and have the uniform distribution on the unit cube $[0,1]^d$. For each n, let $I = I_n$ be chosen independently and uniformly from the set $[n] = \{1, 2, \ldots, n\}$, and let*

$$\mathcal{N}_n^* = \{n^{1/d}(X_i - X_I) : 1 \leq i \leq n\}.$$

We have the following three limits:

(5.15) $$\mathcal{N}_n^* \xrightarrow{d} \mathcal{N}^0,$$

(5.16) $$\deg(X_I, \mathcal{N}_n) \to D = \deg(0, \mathcal{G}(\mathcal{N}^0)) \quad \text{in } L^1,$$

and

(5.17) $$\mathcal{T}(\mathcal{N}_n^*) \xrightarrow{d} \mathcal{G}(\mathcal{N}^0).$$

Proof. The first assertion, (5.15), is classical and will not be proved here. We will first establish (5.16). By (5.15), we can assume that we have versions of the processes $\mathcal{N}_n^*$ and $\mathcal{N}^0$ (and with the primes suppressed) such that with probability one, we have $\mathcal{N}_n^* \to \mathcal{N}^0$ in the sense of set convergence. Also, if we write $\mathcal{N}_n^* = \{x_{n,i}\}$, there is no loss of generality in assuming that $x_{n,n} = X_I$. Next, we collect some information about $D_{n,n} = \deg(x_{n,n}, \mathcal{N}_n^*)$ and its relation to $D = \deg(0, \mathcal{G}(\mathcal{N}^0))$ in these versions.

By Lemma 5.4.1, we have $ED = 2$, and by the exchangeability of $\{x_{n,i}\}$ together with the fact that the average degree in any spanning tree on n vertices is $2(n-1)/n$, we have $ED_{n,n} = 2(n-1)/n$. This gives us trivially that $ED_{n,n} \to ED$. Since by the semicontinuity inequality (5.9) of Lemma 5.3.1, we already have that

$$\liminf_{n \to \infty} D_{n,n} \leq D \quad \text{a.s.},$$

we can conclude by Lemma 5.3.2 that we have $D_{n,n} \to D$ in L^1, thus establishing (5.16).

Now we deal with (5.17). If we write $\mathcal{N}^0 = \{x_i\}$, then by Lemma 5.3.1 we can prove (5.17) if we show that with probability one, we have for all L such that $S \cap \partial C_L = \emptyset$ that

(5.18) $$\limsup_n \sum_{x_{n,i} \in C_L} \deg(x_{n,i}, G(S^n)) \leq \sum_{x_i \in C_L} \deg(x_i, G(S)).$$

As before, our approach is to use expectations and semicontinuity. To keep the notation tidy, we let

$$Y_n(L) = \sum_{x_{n,i} \in C_L} \deg(x_{n,i}, G(S^n)) \quad \text{and} \quad Y(L) = \sum_{x_i \in C_L} \deg(x_i, G(S)).$$

To show $E|Y_n(L) - Y(L)| \to 0$, we just need to prove that

$$\liminf_{n \to \infty} Y_n(L) \leq Y(L) \quad \text{a.s. and} \quad \lim_{n \to \infty} EY_n(L) = EY(L).$$

We first consider the expectation EY. By the homogeneity of the Poisson process, we have an exact formula

(5.19) $$EY = ED \operatorname{card} \mathcal{N}^0 \cap C_L,$$

and by exchangeability, we have a finite-sample analogue

(5.20) $$EY_n = ED_n \operatorname{card} \mathcal{N}_n^* \cap C_L.$$

Now we need to argue that $EY_n \to EY$ or the equivalent proposition that for $Z_n = D_n \operatorname{card} \mathcal{N}_n^* \cap C_L$ and $Z = D \operatorname{card} \mathcal{N}^0 \cap C_L$, we have $EZ_n \to EZ$. The argument is not hard, but some twisting is required because of the several types of convergence that are floating around. We begin by collecting a few facts. First, our choice of versions guarantees that

$$\operatorname{card} \mathcal{N}_n^* \cap C_L \to \operatorname{card} \mathcal{N}^0 \cap C_L \quad \text{a.s.}$$

Second, we just saw that $D_{n,n} \to D$ in L^1. Also, by direct calculations that we can safely skip, we have that for each fixed L that the collection $\{\operatorname{card} \mathcal{N}_n^* \cap C_L : 1 \leq n < \infty\}$ is uniformly bounded in L^2. The variables $D_{n,n}$ are uniformly bounded by b_d, so we further see that $\{Z_n : 1 \leq n < \infty\}$ is uniformly integrable.

Now, given any subsequence of $n = 1, 2, \ldots$, by our first two facts, we can find a further subsequence n_k such that Z_{n_k} converges almost surely to Z. By the uniform integrability of $\{Z_n : 1 \leq n \leq \infty\}$, we therefore find $EZ_{n_k} \to EZ$. By the arbitrariness of our initial subsequence, we therefore find that $EZ_n \to EZ$. By the identities (5.19) and (5.20), we then see that $EY_n(L) \to EY(L)$, so by Lemma 5.3.2 we find for each L that $Y_n(L) \to Y(L)$ in L^1. Finally, by the usual diagonal argument for the rationals, we can find a subsequence $\{n_k\}$ and set Ω_0 of probability one such that for all rational L and all $\omega \in \Omega_0$, we have $Y_n(L)(\omega) \to Y(L)(\omega)$. This gives us condition (5.10) of Lemma 5.3.1, so we have (5.17), completing the proof of the lemma.

Final steps for Theorems 5.2.1 and 5.2.2.

The proof of Theorem 5.2.2 is already complete by Lemma 5.4.1 and by equation (5.16) of Lemma 5.5.1, so we only need to show Theorem 5.2.1. If we write $\{L_{n,j}(X_i) : j = 1, 2, \ldots\}$ for the set of lengths of the edges of $\mathcal{N}_n$ that are incident at X_i, then since every edge meets two vertices, we have

$$Q_n \equiv \sum_{i=1}^{n-1} |e_i|^d = \sum_{i=1}^{n} \frac{1}{2} \sum_j L_{n,j}^d(X_i).$$

By taking expectations and using exchangeability, we then find

$$EQ_n = \frac{n}{2} E \sum_j L_{n,j}^d(X_I),$$

where X_I is a vertex of $\mathcal{N}_n$ that is chosen independently of $\mathcal{N}_n$ with the uniform distribution. The tree convergence (5.17) of Lemma 5.5.1 gives us

$$(n^{1/d} L_{n,j}(X_I); j \geq 1) \xrightarrow{d} (L_j; j \geq 1),$$

so the convergence of the expectations $EQ_n \to l_d < \infty$ will follow from uniform integrability given by the first assertion of the next lemma.

LEMMA 5.5.2.

(5.21) $$\left\{ n \sum_j L_{n,j}^d(X_I) : n \geq 1 \right\} \text{ is uniformly integrable}$$

and

(5.22) $$\left\{ nE \sum_i |e_i|^{2d} : n \geq 1 \right\} \text{ is uniformly bounded.}$$

Proof. We will first show that (5.22) implies (5.21), and then we will establish (5.22). We again recall that an MST in dimension d cannot have any degree greater than a constant $b_d < \infty$, so by Schwarz's inequality we have

(5.23) $$\left(n \sum_j L_{n,j}^d(X_I) \right)^2 \leq b_d n^2 \sum_j L_{n,j}^{2d}(X_I).$$

Now if we calculate the expectation of the right-hand side of (5.23) by first conditioning on $\mathcal{N}_n$, we see that it equals

(5.24) $$2nb_d \sum_{e \in M} |e|^{2d},$$

where we have set $M \equiv \text{MST}(\mathcal{N}_n)$. Thus we find that boundedness of the expectations in (5.22) will imply the L^2 boundedness of $\{n \sum_j L_{n,j}^d(X_I) : n \geq 1\}$, which is more than we need for the uniform integrability (5.21).

The proof of (5.22) will offer another illustration of the analytical use of the space-filling curves introduced in the Chapter 3. We let $\phi : [0,1] \to [0,1]^d$ be any measure-preserving transformation which is a surjection from $[0,1]$ onto $[0,1]^d$ and which is Lipschitz of order $1/d$,

$$|\phi(x) - \phi(y)| \leq c|x - y|^{1/d} \quad \text{for all } x, y \in [0,1]^d.$$

As in the Chapter 3, we note that if the random variables $\{U_i\}$ are independent and uniformly distributed on $[0,1]$, then the variables $X_i = \phi(U_i)$ are independent and uniformly distributed on $[0,1]^d$. Also, if one takes the edge costs to be $|e|^{2d}$ instead of $|e|$, the set of edges of the MST does not change, so we see that the spanning tree given by the space-filling curve heuristic provides a bound on the right side of (5.24):

$$2nb_d \sum_{e \in M} e^{2d} \leq 2nb_d \sum_{1 \leq i < n} |\phi(U_{(i)}) - \phi(U_{(i+1)})|^{2d},$$

where the $\{U_{(i)} : 1 \leq i \leq n\}$'s are the order statistics of the $\{U_i\}$'s. By the Lipschitz property of ϕ, the last term is bounded by

$$cnb_d \sum_{1 \leq i < n} |U_{(i)} - U_{(i+1)}|^2,$$

and by elementary calculus, $E|U_{(i)} - U_{(i+1)}|^2 = O(1/n^2)$. This proves that the sum (5.24) has expectation that is bounded independently of n. Thus the left side of (5.23) also has uniformly bounded expectation. This is just what is needed to prove (5.22), the second assertion of the lemma.

With the last lemma, we have completed the proof that

(5.25) $$Q_n \equiv \sum_{i=1}^{n-1} |e_i|^d \quad \text{satisfies} \quad EQ_n \to l_d.$$

With (5.25) in hand, the usual subsequence arguments tell us that to show that $Q_n \to l_d$ in L^p for all $p \geq 1$, we only need to show that Q_n is uniformly bounded. We have already established such a bound for $d = 2$ in Chapter 3 by the use of the Lipschitz-$\frac{1}{2}$ space-filling curves, and the same proof can be extended to all $d \geq 2$. We will give an independent proof that illustrates some of the geometry of the MST in $\mathbb{R}^d$.

Lens geometry for the MST.

For any $x \in \mathbb{R}^d$ and $r > 0$, we let $D(x,r) = \{z : |x - z| < r\}$, and for any $x, y \in \mathbb{R}^d$, we let $L(x,y) = D(x, |x-y|) \cap D(y, |x-y|)$. The set $L(x,y)$ and sets that are geometrically similar to $L(x,y)$ are commonly called lenses. By $L^\alpha(x,y)$ we denote the region similar to $L(x,y)$ with center $(x+y)/2$ but dilated by α. Formally,

$$L^\alpha(x,y) = \{z : (x+y)/2 + \alpha^{-1}(z - (x+y)/2) \in L(x,y).$$

To confirm the notation, one should note that $L(x,y)$ is contained in the ball of radius $r = |x - y|$ with center $(x+y)/2$ and $L(x,y)$ contains the ball of radius $r/2$ with center $(x+y)/2$. Similarly, $L^\alpha(x,y)$ is contained in the ball of radius $\alpha r = \alpha|x-y|$ with center $(x+y)/2$ and $L^\alpha(x,y)$ contains the ball of radius $\alpha r/2$ with center $(x+y)/2$. A well-known consequence of the triangle inequality is that for the MST of n points $S \cap \{x_1, x_2, \ldots, x_n\} \subset \mathbb{R}^d$, one has $S \cap L(x_i, x_j) = \emptyset$ for all $1 \leq i < j \leq n$. The following lemma gives a very useful property of the sets $L^\alpha(x_i, x_j)$.

LEMMA 5.5.3. *For the MST of n points $S = \{x_1, x_2, \ldots, x_n\} \subset \mathbb{R}^d$, one has that the sets $L^\alpha(x_i, x_j)$ for all $1 \leq i < j \leq n$ are all disjoint provided that $\alpha < \frac{1}{2}(2 + \sqrt{3})^{-1}$.*

Proof. Suppose that z is in both $L^\alpha(a,b)$ and $L^\alpha(a',b')$, where $|a' - b'| = r' \leq |a - b| = r$. If we write $c = (a-b)/2$ and $c' = (a'-b')/2$, then we have

$$|z - c'| \leq |z - c| \leq \alpha r \sqrt{3}/2 \quad \text{and} \quad |c' - a'| \leq \frac{1}{2}\alpha r.$$

Thus by the triangle inequality and the condition on α, we see that

$$|c - a'| \leq \alpha(2 + \sqrt{3})r < r/2.$$

This condition implies that $a' \in L^\alpha(a,b)$, and this is not possible because, as we noted before, $S \cap L(x_i, x_j) = \emptyset$ for all $1 \leq i < j \leq n$.

The real punchline of the last lemma is that for each $d \geq 2$, there is a constant C_d such that for all n the edges e_i of the MST of $S \cap \{x_1, x_2, \ldots, x_n\} \subset [0,1]^d$ satisfy

(5.26) $$\sum_{i=1}^{n-1} |e_i|^d \leq C_d,$$

as we can check with a little geometry. For any edge (x_i, x_j) of the MST of $S \cap \{x_1, x_2, \ldots, x_n\} \subset [0,1]^d$, set $L(x_i, x_j)$ is (too generously) contained in the box $B_d = [-1,2]^d$. By the fact that the $L^\alpha(x_i, x_j)$'s are disjoint for $(x_i, x_j) \in \text{MST}(S)$,

$$\text{vol}_d \bigcup_{(i,j) \in \text{MST}} L^\alpha(x_i, x_j) \leq \sum_{(i,j) \in \text{MST}} \text{vol}_d L^\alpha(x_i, x_j) \leq \text{vol}_d B_d.$$

The bound (5.26) then follows from the fact that there is a constant β_d such that for all x and y, the volume of $L(x,y)$ is equal to $\beta_d |x-y|^d$. In (5.26), we can take C_d equal to $\text{vol}_d B_d / (\alpha^d \beta_d)$, though this is quite crude.

5.6. Central limit theory.

The possibility of a central limit theorem for the traveling-salesman problem was already suggested in Beardwood, Halton, and Hammersley (1959), but until relatively recently the prospects for a CLT for any subadditive Euclidean functional seemed remote. The earliest work on such a CLT was done by Ramey (1983), who showed that if one could prove a certain hypothesis in continuum percolation theory, then one could prove a CLT for the minimal spanning tree. There were many useful ideas in Ramey (1983), but a proof of the required hypothesis remained elusive. The next important step in the central limit theory for subadditive Euclidean functionals was made in Avram and Bertsimas (1993). We will develop their contribution over the next few sections, but before beginning that development, we should report on definitive new progress on the CLT for the minimal spanning tree.

First, in dimension two, Alexander (1995a) proved the conjecture of Aldous and Steele (1992) on the nature of the spanning tree for the Poisson process. In particular, Alexander (1995a) proved that the forest constructed by Aldous and Steele (1992) is indeed a tree, thus nailing down the core reason for the effectiveness of the construction. Further, Alexander (1995a) obtained useful information about the structure of the spanning tree for the Poisson process by proving that from each point of the tree there is a unique path to infinity. One of the consequences of this structural result is an alternative characterization of the limiting constant given in Theorem 5.2.1.

In that theorem, we found that the dth power of the edge lengths of the MST of a uniform sample from $[0,1]^d$ converged to a constant,

$$\sum_{e \in \text{MST}(S^n)} |e|^2 \to l_d \quad \text{in } L^p \quad \text{as } n \to \infty.$$

Moreover, if we take the lengths $\{L_i\}$ of the edges of the Poisson minimal forest that are incident to the point 0, then we can identify the limit constant by the quantity

(5.27) $$l_d \equiv \frac{1}{2} \sum_i E L_i^d < \infty.$$

Now given Alexander's structural results, we have a much more intuitive interpretation of the limiting constant l_d when $d = 2$. Specifically, if we take $\hat{e}$ to be the unique edge that is incident to 0 and that is on the path to infinity, then we can simply take l_2 to be $E(\hat{e}^2)$.

Finally, Alexander (1996) proved a central limit theorem for the Poissonized MST in $d = 2$. The outline of Alexander's approach had been used by Ramey (1982) to obtain a CLT for the MST under the assumption of a technical anzatz.

THEOREM 5.6.1 (Alexander's CLT for the MST). *Let the independent random variables $\{X_i : 1 \leq i < \infty\}$ have the uniform distribution in $[0,1]^2$, and let $N(t)$ be an independent Poisson counting process with unit rate. If $M(t) \equiv M(X_1, X_2, \ldots, X_{N(t)})$ denotes the length of the minimal spanning tree of the set of points $\{X_1, X_2, \ldots, X_{N(t)}\}$, then $\inf_{t:t\geq 1} \operatorname{Var}(M(t)) > 0$ and for $t \to \infty$ we have convergence in distribution of the standardized MST length to a standard normal:*

$$\frac{M(t) - EM(t)}{\sqrt{\operatorname{Var} M(t)}} \Rightarrow N(0,1).$$

Shortly after the developments provided by Alexander (1996), a central limit theorem was proved by Kesten and Lee (1996) that went much farther. In particular, the theorem of Kesten and Lee (1996) covered all dimensions, made the normalizing variance more explicit, and covered the case of power-weighted edges.

THEOREM 5.6.2 (Kesten and Lee's CLT for the MST). *Let $\{X_i : 1 \leq i < \infty\}$ be independent random variables with the uniform distribution on $[0,1]^d$. For any $\alpha > 0$, if we let*

$$M(X_1, X_2, \ldots, X_n; \alpha)$$

$$= \min\left\{\sum_{e \in T} |e|^\alpha : T \text{ is a spanning tree of } \{X_1, X_2, \ldots, X_n\}\right\},$$

then we have convergence in distribution to a normal

$$n^{-(d-2\alpha)/2d}\{M(X_1, X_2, \ldots, X_n; \alpha) - EM(X_1, X_2, \ldots, X_n; \alpha)\} \Rightarrow N(0, \sigma^2_{\alpha,d}),$$

where $\sigma^2_{\alpha,d}$ is a strictly positive constant.

The proof of Kesten and Lee (1996) was structured around the use of the martingale central limit theorem. This approach reduces the proof of the CLT to the proof of a weak law of large numbers for a certain sequence of conditional variances. The reduction is quite useful, but the proof of the required weak law of large numbers still turns out to be more complicated than we can give here.

The limitations of the Kesten and Lee (1996) argument seem to come mostly from the extent to which it rests on the characterization of the MST in terms of the greedy algorithm. Such a characterization is not available in the TSP or in the minimal-matching problems, so we still seem to be far away from a CLT for these problems. Also, as the argument now stands, we do not know if it can be modified to provide any information about the rate of convergence to the normal law.

5.7. Conditioning method for independence.

Perhaps the most persistent source of bedevilment in applied probability is the approximate independence that one can see but cannot hold. The central limit theorems of Alexander (1996) and Kesten and Lee (1996) provide telling illustrations of how hard one must work sometimes in order to extract what is needed from random variables that seem intuitively to have all the independence that one might wish. The main goal of this section is to develop a method due to Avram and Bertsimas (1993) that often gives us a direct handle on the independence in geometrical problems that exhibit elusive independence. The method is simple, powerful, and widely applicable. For many problems of the probability theory of geometric combinatorial optimization, the Avram–Bertsimas technique should be the first method to be tried since even when the technique is not successful, the issues that are brought forth can be informative. The main limitation of the Avram–Bertsimas technique is that it applies only to problems that have a kind of combinatorial localizability. The TSP, MST, and minimal matching apparently do not have such localizability, but many related problems do.

The idea behind the Avram–Bertsimas technique is that one can often find an event such that by conditioning on the occurrence of that event, the random variables that were approximately independent in the unconditioned model become genuinely independent under the conditional model. We will illustrate the idea by establishing a strong version of the central limit theorem for Z_n of the nearest-link problem. The main benefit of this problem is that it gives us an example where the appropriate localizability can be obtained with very little effort.

To describe the simplest version of nearest-neighbor-link problem, we again take independent random variables $\{X_i : 1 \leq i < \infty\}$ with the uniform distribution on $[0,1]^2$ and consider the set of points $\mathcal{N}_n \equiv \{X_i : 1 \leq i \leq N(n)\} \subset [0,1]^2$, where $N(t)$ is an independent Poisson counting process with unit rate. The quantity of interest Z_n is the sum of the distances between the nearest-neighbor pairs in $\mathcal{N}_n$, where y is called a *nearest neighbor* of x in $\mathcal{N}_n$ if the open ball of radius $|x-y|$ with center x contains no point of $\mathcal{N}_n$ except x.

To build some intuition about Z_n, we recall some elementary facts. In our model, each point of $\mathcal{N}_n$ has a unique nearest neighbor. Still, we can have y as the nearest neighbor of x while x is not the nearest neighbor of y. In fact, we even expect a positive fraction of elements of $\mathcal{N}_n$ to exhibit this behavior. Also, if t is large and x and y are not close, then one surely suspects that the lengths of the distances from x and y to their respective nearest neighbors should be nearly independent. This is indeed the case, but the approximate independence is not easy to articulate or to use. Only by dint of considerable effort could Bickel and Breiman (1983) exploit enough independence to be able to prove that Z_t satisfies a central limit theorem. The Avram–Bertsimas technique turns out to give us a quicker path to a more precise result.

To spell out our conditioning, we first let $C = \{C_{ij} : 1 \leq i \leq m, 1 \leq j \leq m\}$ denote the set of m^2 subsquares of $[0,1]^2$ given by $C_{ij} = [(i-1)/m, i/m] \times [(j-1)/m, j/m]$. Also, we will take $m = \lfloor n^{(1-\delta)/2} \rfloor$, where $\frac{1}{2} > \delta > 0$ will be chosen

later. We then write
$$Z_n = \sum_{C_{ij} \in C} L(C_{ij}),$$
where $L(C_{ij})$ denotes the sum of the length of the nearest-neighbor links that are contained in C_{ij} together with that part of the lengths of the links that have just one point of $\mathcal{N}_n$ in C_{ij}. We now take

(5.28) $\quad A_n = \{1 \leq \text{card } C_{ij} \cap \mathcal{N}_n \leq \lceil en/m^2 \rceil \text{ for all } 1 \leq i,j \leq m\}.$

Since $\text{card } C_{ij} \cap \mathcal{N}_n$ has the Poisson distribution with parameter $\lambda = n/m^2$, standard estimates give us that

(5.29) $\quad P\{1 \leq \text{card } C_{ij} \cap \mathcal{N}_n \leq \lceil en/m^2 \rceil\} \geq 1 - 2e^{-\lambda},$

so independence and our choice $m = \lfloor n^{(1-\delta)/2} \rfloor$ give us

(5.30) $\quad (1 - 2e^{-\lambda})^{m^2} \leq P(A_n) \leq 1,\quad \text{and hence}\quad 1 - P(A_n) = O(ne^{-n^\delta}).$

We now introduce a new probability measure
$$\hat{P}(B) \equiv P(B \cap A_n)/P(A_n),$$
and we note that since $P(A_n)$ is close to 1, the new measure $\hat{P}$ is approximately equal to P. The benefit of swapping $\hat{P}$ for P is that there are many geometrically important events that are perfectly independent under $\hat{P}$ but which are only approximately independent under P.

We will shortly need the fact that all of the moments of Z_n under P are close to the corresponding moments under $\hat{P}$. This is easy to prove. Even if we use the crude bound $Z_n \leq 2\sqrt{d}\,\text{card}\,\mathcal{N}_n$, inequality (5.30) is strong enough to give us for all $1 \leq k < \infty$ that

(5.31) $\quad E_P Z_n^k - E_{\hat{P}} Z_n^k = o(e^{-n^\delta/2}).$

Second look at $\hat{P}$.

A useful alternative description of the measure $\hat{P}$ can be given by providing an explicit construction of the point process with the law $\hat{P}$. To this end, we first consider m^2 independent random variables N_{ij} with the doubly truncated Poisson distribution with parameter $\lambda = n/m^2$, which is to say, the event $E_k = \{N_{ij} = k\}$ has probability proportional to $\lambda^k e^{-\lambda}/k!$ for all integers $1 \leq k \leq \lceil en/m^2 \rceil$ and E_k has probability 0 for all other integers. One then constructs the random point set $\mathcal{N}$ in $[0,1]^2$ by choosing N_{ij} independent and uniformly distributed points out of C_{ij} for each $1 \leq i,j \leq m$. The benefit of this construction over the direct definition of $\hat{P}$ is that we can now easily see that if S and S' are unions of disjoint subsets of the C_{ij}, then the point processes $\mathcal{N} \cap S$ and $\mathcal{N} \cap S'$ are independent.

We now introduce a distance between pairs of subsets C, our set of small cubes, by taking
$$\gamma(A,B) = \min\{\max(|i - i'|, |j - j'|) : C_{ij} \subset A \text{ and } C_{i',j'} \subset B\}.$$

Perhaps the central observation of Avram and Bertsimas (1993) is the following simple proposition together with its natural extensions.

PROPOSITION 5.7.1. *For any pair of sets of subcubes A and B with $\gamma(A,B) \geq 4$, the sigma-fields $\sigma\{L(C_{ij}) : C_{ij} \in A\}$ and $\sigma\{L(C_{ij}) : C_{ij} \in B\}$ are independent with respect to the probability measure $\hat{P}$.*

Proof. By construction of the point process $\mathcal{N}$ with the law $\hat{P}$, each subcube C_{ij} always contains at least one point of $\mathcal{N}$. By easy geometry, we then see that for any $x \in C_{ij}$ the nearest neighbor of x in $\mathcal{N}$ is in a C_{st} such that $\gamma(C_{ij}, C_{st}) \leq 2$. Moreover, the random variable $L(C_{ij})$ is measurable with respect to the sigma-field generated by $\{\mathcal{N} \cap C_{st} : \gamma(C_{st}, C_{ij}) \leq 2\}$. By applying the same observation to each $C_{ij} \in A$ and to $C_{ij} \in b$, we find

$$\sigma\{L(C_{ij}) : C_{ij} \in A\} \subset \sigma\{\mathcal{N} \cap C_{st} : \gamma(C_{st}, A) \leq 2\}$$

and

$$\sigma\{L(C_{ij}) : C_{ij} \in B\} \subset \sigma\{\mathcal{N} \cap C_{st} : \gamma(C_{st}, B) \leq 2\}.$$

By hypothesis, $\gamma(A,B) \geq 4$, so the sigma-fields on the right side of the last two inclusions are independent, and the lemma is proved.

From the proof of the proposition, we see that if we had taken

(5.32) $$A'_n = \{1 \leq \operatorname{card} C_{ij} \cap \mathcal{N}_n \text{ for all } 1 \leq i, j \leq m\}$$

instead of A_n, then we would still have obtained the desired independence. The additional constraint that $\operatorname{card} C_{ij} \cap \mathcal{N}_n \leq \lceil en/m^2 \rceil$ was imposed on A_n only because the constraint will be useful later. Thus, just as adenoids are often taken out along with tonsils, we have elected to operate only once on our point process.

5.8. Dependency graphs and the CLT.

There are many situations where independence relationships between random variables can be expressed most clearly with graph-theoretical language, and one particularly useful construction for that purpose is the dependency graph. If $\mathcal{X} = \{X_v : v \in V\}$ is a collection of random variables, a graph $G = (V, E)$ is called a *dependency graph* for $\mathcal{X}$ if $G = (V, E)$ has the property that the sigma-fields $\sigma\{X_v : v \in A\}$ and $\sigma\{X_v : v \in B\}$ are independent whenever there is no edge in G between A and B. The notion of dependency graphs has not been widely used by probabilists, but, ever since the development of the Lovász local lemma (Erdös and Lovász (1975)), dependency graphs have been of fundamental use for proving existence of combinatorial objects by the probabilistic method (cf. Alon, Spencer, and Erdös (1992, pp. 53–69)). For our immediate purposes, the main result that we will use from the probability theory of dependency graphs is a central limit theorem of Baldi and Rinnot (1989).

THEOREM 5.8.1 (Baldi and Rinott (1989)). *Suppose that for all $n \geq 1$ the collection of random variables $\{X_{vn} : v \in V_n\}$ has dependency graph $G_n = (V_n, E_n)$ and G maximum degree D_n. If $S_n = \sum_{v \in V_n} X_{vn}$ and F_n denotes the distribution function of $(S_n - ES_n)/\sigma_n$, where $\operatorname{Var}(S_n) \equiv \sigma_n^2 < \infty$, then we have*

$$|F_n(x) - \Phi(x)| \leq 32(1 + \sqrt{6})(\operatorname{card} V_n)^{1/2} D_n (B_n/\sigma_n)^{3/2}$$

provided that the constant B_n satisfies

$$\max_{v \in V_n} |X_{vn}| \leq B_n.$$

Baldi and Rinott (1989) obtained this result as a corollary to a more general theorem where, instead of boundedness of the summands, one only requires the boundedness of a certain fourth moment. The more general result of Baldi and Rinott is itself obtained by an appropriate specialization of the general dependent central limit theorem of Stein (1986, p. 110). We will apply Theorem 5.8.1 shortly, but first we need some information about the variance of the sum of the lengths of the nearest-neighbor links under our two probability models P and $\hat{P}$.

LEMMA 5.8.1. *There is a constant $\kappa > 0$ and an integer $2 \leq n_0 < \infty$ such that for all $n \geq n_0$, we have*

(5.33) $$\operatorname{Var}_P(Z_n) \geq \kappa$$

and

(5.34) $$\operatorname{Var}_{\hat{P}}(Z_n) \geq \kappa.$$

Proof. There is a vague but useful principle for stationary processes that says "if something can happen, it will happen regularly." To focus the principle on (5.33), we first partition the square $[0,1]^2$ into q^2 equal subsquares C_i, where $q = \lceil \sqrt{n} \rceil$ and $1 \leq i \leq q^2$. We then fix $\epsilon = (13q)^{-1}$ and partition each C_i into 13×13 subsubsquares of side length ϵ as indicated in Figure 5.1. We let $S(i,1)$ be the subsubsquare in the center of C_i and we let $S(i,2)$ be the subsubsquare that is two subsubsquares to the right of $S(i,1)$. For each subsquare C_i, we consider the event $U(C_i)$ of small—but positive—probability that $C_i \cap \mathcal{N}$ satisfies the following conditions:

- both $S(i,1)$ and $S(i,2)$ contain exactly one point of $\mathcal{N}$,

- each of the 48 "boundary subsubsquares" of edge length ϵ that are interior to C_i and share an edge with the boundary of C_i contains exactly one point of $\mathcal{N}$, and

- the rest of the square C_i contains no points of $\mathcal{N}$.

We now consider a typical square C_i that satisfies the conditions just given. We let X_i and y_i denote the unique points in $S(i,1)$ and $S(i,2)$; the reason for the differing capitalizations will be evident shortly.

We let S denote the set of the indices of the C_i for which $U(C_i)$ occurs, and, parallel to our earlier analysis, we let $Z_n(C_i)$ denote that part of the nearest-neighbor graph that is in the cube C_i. Finally, we next introduce a sigma algebra that keeps track of everything except the location of X_i in the C_i for which $U(C_i)$ occurs. More formally, we let $\mathcal{F}$ denote the sigma algebra generated by S and

$$\mathcal{N} \cap \left(\bigcup_{i \in S} S(i,1) \right)^c.$$

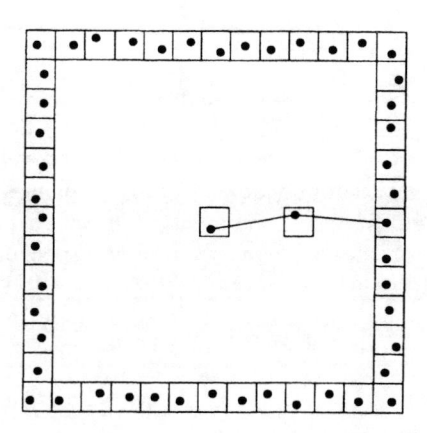

FIG. 5.1. *Square decomposition for variance lower bound 5.34.*

We can now happily calculate

$$\operatorname{Var}_P(Z_n) = \operatorname{Var}_P(E_P(Z_n|\mathcal{F})) + E_P \operatorname{Var}_P(Z_n|\mathcal{F})$$

$$\geq E_P \operatorname{Var}_P \left(\sum_{i \in S} Z_n(C_i) + \sum_{i \notin S} Z_n(C_i) \,|\, \mathcal{F} \right)$$

$$= E_P \operatorname{Var}_P \left(\sum_{i \in S} Z_n(C_i) \,|\, \mathcal{F} \right)$$

$$= E_P \operatorname{Var}_P \left(\sum_{i \in S} |y_i - X_i| \,|\, \mathcal{F} \right)$$

$$\geq n^{-1} \tau E_P \left(\sum_{i \in S} 1 \right),$$

where in the last step we used the independence of $|y_i - X_i|$ given $\mathcal{F}$ and the fact that there is a constant $\tau > 0$ such that for all n we have $\operatorname{Var}_P(|y_i - X_i| \,|\, \mathcal{F}) \equiv v_n(y_i) \geq n^{-1}\tau$ for all $y_i \in S(i,2)$. There is a $p > 0$ and an n_0 such that for all $n \geq n_0$ we have $P(C_i \in S) \geq p$, so we see that a lower bound of $p\tau$ can be used in (5.33). The lower bound in (5.34) follows from (5.33) and the moment bound (5.31).

We now have all the tools needed to prove a central limit theorem and a rate of convergence. In order to keep our estimates as simple as possible, the rate that is proved here is somewhat slower than the rate given in Avram and Bertsimas (1993).

THEOREM 5.8.2. *Let the independent random variables $\{X_i : 1 \leq i < \infty\}$ have the uniform distribution in $[0,1]^2$, and let $N(t)$ be an independent Poisson counting process with unit rate. If Z_n denotes the sum of the distances between the nearest-neighbor pairs of $\mathcal{N}_n = \{X_i : 1 \leq i \leq N(n)\}$, then for all $\epsilon > 0$,*

$$(5.35) \qquad P\left((Z_n - EZ_n)/\sqrt{\operatorname{Var} Z_n} \leq x\right) - \Phi(x) = O(n^{-1/4+\epsilon}).$$

Proof. We first prove a version of the theorem under the conditional probability measure $\hat{P}$. As before, we take $C_{ij} = [(i-1)/m, i/m] \times [(j-1)/m, j/m]$, where $m = \lfloor n^{(1-\delta)/2} \rfloor$, and we put $\hat{P}(B) \equiv P(B \cap A_n)/P(A_n)$, where A_n is given by

(5.36) $\quad A_n = \{1 \leq \text{card } C_{ij} \cap S_n \leq \lceil en/m^2 \rceil \text{ for all } 1 \leq i, j \leq m\}.$

We write
$$Z_n = \sum_{C_{ij} \in C} L(C_{ij}),$$
where $L(C_{ij})$ denotes the contribution to Z_n from links that are partly or wholly in C_{ij}.

Under the probability model $\hat{P}$, the dependency graph $G_n = (V_n, E_n)$ for the collection of variables $V_n = \{L(C_{ij}) : 1 \leq i, j \leq m\}$ will have no edge between $L(C_{ij})$ and $L(C_{i'j'})$ if $\gamma(C_{ij}, C_{i'j'}) \geq 4$. Hence we see that the maximum degree in G_n is bounded by 25 for all n.

Also, under the model $\hat{P}$, we know that $L(C_{ij})$ is bounded above by
$$2\sqrt{d}\, m^{-1} \text{card } \mathcal{N}_n \cap C_{ij} \leq 2m^{-1} en/m^2,$$
so we have
$$B_n = O(m^{-1} n^\delta) = O(n^{-1/2 + (3/2)\delta}).$$

The bounds on D_n, B_n, and σ_n tell us that $((\text{card } V_n)^{\frac{1}{2}} D_n (B_n/\sigma_n)^{\frac{3}{2}}) = O(n^{\delta + \frac{1}{4}})$, so Theorem 5.8.1 gives us our CLT for Z_n under the $\hat{P}$ model:
$$\hat{P}\left(\frac{Z_n - E_{\hat{P}} Z_n}{\sqrt{\text{Var}_{\hat{P}} Z_n}} \leq x\right) - \Phi(x) = O(n^{\delta + 1/4}).$$

To complete the proof, we just need to show that we can pass from the measure $\hat{P}$ to our original P without greatly disturbing the distribution functions. We first note by the definition of $\hat{P}$ that
$$P(A_n)\hat{P}(Z_n \leq x) = P(Z_n \leq x, A_n) \leq P(Z_n \leq x)$$
and
$$P(A_n)\hat{P}(Z_n \leq x) = P(Z_n \leq x, A_n) \leq P(Z_n \leq x) + P(A_n^c);$$
so, by the bound $P(A_n) - 1 = O(e^{-n^\delta})$ that we obtained in (5.30), we find

(5.37) $\quad P(Z_n \leq x) - \hat{P}(Z_n \leq x) = O(e^{-n^\delta}) \quad \text{for all } x.$

Now to bring in the normalized sums, we recall that from (5.31) we have for all k that
$$E_P Z_n^k - E_{\hat{P}} Z_n^k = o(e^{-n^\delta/2}),$$
so we also have $\text{Var}_P(Z_n) - \text{Var}_{\hat{P}}(Z_n) = o(e^{-n^\delta/2})$.

The bounds on $E_P Z_n - E_{\hat{P}} Z_n$ and $\text{Var}_P(Z_n) - \text{Var}_{\hat{P}}(Z_n)$ tell us that we can change x just a little to get an x' that will let us take the hats off inside of $\hat{P}$. More precisely, there is an $x' = x'(x)$ such that $|x - x'| = o(e^{-n^\delta/2})$ and
$$\hat{P}(Z_n \leq E_P Z_n + x\sqrt{\text{Var}_P Z_n}) = \hat{P}(Z_n \leq E_{\hat{P}} Z_n + x'\sqrt{\text{Var}_{\hat{P}} Z_n}).$$
We have from (5.37) and the bound on $|x - x'|$ that
$$\begin{aligned} P\left(Z_n \leq E_P Z_n + x\sqrt{\text{Var}_P Z_n}\right) &= \hat{P}\left(Z_n \leq E_P Z_n + x\sqrt{\text{Var}_P Z_n}\right) + O\left(e^{-n^\delta}\right) \\ &= \hat{P}\left(Z_n \leq E_{\hat{P}} Z_n + x'\sqrt{\text{Var}_{\hat{P}} Z_n}\right) + O\left(e^{-n^\delta}\right) \\ &= \Phi(x') + O(n^{\delta-1/4}) + O(e^{-n^\delta}). \end{aligned}$$
Since $|\Phi(x') - \Phi(x)| \leq |x - x'| = o(e^{-n^\delta/2})$, we have the bottom-line equality
$$P\left(Z_n \leq E_P Z_n + x\sqrt{\text{Var}_P Z_n}\right) - \Phi(x) = O(n^{\delta-1/4}),$$
just as required.

Additional functionals.

The analysis given for the nearest-neighbor graph can be applied with minor changes to many other problems. The most immediate extension is to the graph where there is an edge from x to each of its k nearest neighbors. Also, there is no need in this analysis to restrict attention just to edge lengths; one can just as well look at functions of the edge lengths.

A more striking application of the Avram–Bertsimas method is to the Voronoi tessellation and the Delauney triangulation. In these problems, we begin with a finite $S \subset [0,1]^d$, and we associate with each $x \in S$ the set of points of $[0,1]^d$ that are closer to x than to any other point of S. The resulting decomposition of $[0,1]^d$ is called the *Voronoi tessellation* of $[0,1]^d$ by S, and the Delauney triangulation is subsequently defined by the graph on S that has an edge between x and y whenever the Voronoi regions associated with x and y share a face. The same conditioning event that was used to create the measure $\hat{P}$ for the nearest-neighbor graph can be applied to achieve independence in the Delauney-triangulation problem. The dependency graph turns out to have more edges than were needed in the nearest-neighbor problem, but the maximal degree D_n in the dependency graph is still bounded independently of n.

The last application we mention concerns the central limit theorem for the number of extreme points of a random sample from the unit disk. The expectation of the number of extreme points in this and related problems was first investigated by Rényi and Sulanke (1963, 1964), and more recently central limit theorems were obtained by Groeneboom (1988), Hsing (1992), and Heurter (1994). Avram and Bertsimas (1993) show that the conditioning method can be used to provide rate results that sharpen parts of the central limit theory of Groeneboom (1988) and Hsing (1992). So far, the Avram–Bertsimas technique does not seem to have been applied to any functional central limit theorems, though there is no reason to suspect that the ideas would be any less effective in such contexts.

5.9. Additional remarks.

1. The first half of the chapter follows Aldous and Steele (1992) rather closely except in a few points. Theorem 5.2.1 is boosted from convergence in L^2 to convergence in L^p for all $p \geq 1$, and the proof of Lemma 5.5.2 is simplified by the use of space-filling curves. Also, Lemma 5.5.3 has been added both to give the boost to convergence in L^p and to illustrate some basic MST geometry that will be used again in the next chapter.

2. The uniform boundedness of the sum of the dth power of the edges of the MST should not be taken for granted; the corresponding property does not hold for the TSP. Snyder and Steele (1995) proved that there is a constant c such that for any $\{x_1, x_2, \ldots, x_n\} \subset \mathbb{R}^2$ we have

$$\sum_{e \in \text{TSP}} |e|^2 \leq c \log n,$$

but, more to the point, the results of Bern and Epstein (1993) showed that the $O(\log n)$ bound cannot be improved.

3. The view that subadditive methods cannot yield results like Theorem 5.2.1 needs to be modified in light of resent results of Yukich (1995). In particular, Yukich (1995) shows that for independent X_i, $1 \leq i \leq \infty$, with the uniform distribution on $[0,1]^2$, one has

$$E \min_G \sum_{e \in G} |e|^2 \to \beta_G,$$

where the minimum is over the is the set of spanning tours of $\{X_1, X_2, \ldots, X_n\}$. One of the tricky aspects of this result is that it *does not* tell us that the expectation of the sum of the squares of the edges of the usual TSP tour converges. This would require

$$E \sum_{e \in T(n)} |e|^2 \to \beta_G$$

for the random tours $T(n) = T(X_1, X_2, \ldots, X_n)$ with the property that

$$\sum_{e \in T(n)} |e| = \min_G \sum_{e \in G} |e|,$$

where the minimum is over the is the set of spanning tours of $\{X_1, X_2, \ldots, X_n\}$. This distinction does not occur for the MST because one finds the same trees for the MST whether one takes the edge cost to be $|e|$ or $|e|^d$. Thus the analysis of Yukich (1995) does provide an alternative path to the proof of Theorem 5.2.1, though without the benefit of any identification of the limit.

4. The proof of Lemma 5.8.1 follows Avram and Bertsimas (1993) in outline, but some details have been modified in ways that are suggested by a corresponding argument in Kesten and Lee (1995). The proof of Theorem 5.8.2 also follows the plan of Avram and Bertsimas (1993), but more detail has been given on the

relationship of the two measures P and $\hat{P}$. Our apparently more tedious proof is just more attentive to the difference between the normalizing expectations and variances under the two models.

5. Some of the limiting constants first obtained by subadditive methods have been given nonlimiting interpretations by the introduction of the corresponding infinite objects, but the new interpretation does not seem to bring us any closer to the exact determination of the constant values. In this connection, we should note that Avram and Bertsimas (1992) have given a formula for the value of the MST length constant which can be used to obtain a sequence of approximations of the constant value. The progress offered by Avram and Bertsimas (1992) is substantial, but since their formula involves a sequence of integrals of increasing dimension, there still remain many questions concerning the sense to which their formula provides us with a deeper understanding of the limiting constants.

6. Avram and Bertsimas (1992) posed several conjectures concerning the relationship of $\beta_{\mathrm{MST}}(d)$ to corresponding constants under related independent models. Some of these conjectures have been resolved recently by Penrose (1995a).

7. In Steele, Shepp, and Eddy (1987), it is proved that for an independent, continuously distributed sample in $\mathbb{R}^d$, there is an asymptotic frequency $\alpha_{k,d}$ of points of degree k in the MST of the sample. Penrose (1995b) has proved that $\alpha_{k,d} \to \alpha_k$ as $d \to \infty$, where the constants α_k are equal to corresponding constants of an independent model introduced in Aldous (1990).

8. Rhee and Talagrand (personal communication) have recently observed that by a more careful handling of the subsolutions of the power-weighted MST, one can also deal with the critical case $\alpha = d$ by subadditive methods that parallel those of the earlier chapters. Thus the objective method is not the only way of getting this more delicate result, though naturally the objective method also provides considerable related information on the behavior of the MST for random samples.

CHAPTER 6

Talagrand's Isoperimetric Theory

This chapter introduces Talagrand's theory of isoperimetric inequalities for product spaces. Although the theory is a very recent development, it has already had a profound impact on the probability theory of combinatorial optimization. The aim of this chapter is to give the the reader the quickest possible route to understanding how one can apply Talagrand's theorem to problems like those considered in the preceding chapters.

After introducing some of the fundamental concepts of Talagrand's theory, we give three powerful applications of Talagrand's convex-distance inequality. We then give a proof of Talagrand's inequality using the basic inductive technique that is central to the whole theory. Finally, we give several additional illustrations of how the isoperimetric inequalities can be used and how the new methods compare to earlier techniques.

6.1. Talagrand's isoperimetric theory.

A central motivation for our study of Talagrand's isoperimetric theory is that in virtually every case where the bounded-difference method can be applied, Talagrand's methods do at least as well—and in many instances they do dramatically better. Though this should be motivation enough, there is more— in most cases Talagrand's methods apply quite directly and require very little computation.

Still, there is a price to pay. First, there is a modest amount of noncombinatorial machinery that needs to be developed, or at least one needs to recall some elementary facts about sections and projections in products of measure spaces. More to the point, one has to be willing to engage in a certain level of abstraction. Fortunately, the concrete benefits of this investment come quickly and steadily.

Given a probability space (Ω, Σ, μ) and a subset A of the n-fold product space Ω^n, the isoperimetric theory of product spaces provides estimates of the measure of the set of points of Ω^n that are within a specified "distance of A." The theory benefits from taking a broad view of the distance measure, but for the applications in combinatorial optimization, we can go quite far and still focus on just one specific choice.

The measure of distance that we will find most useful is Talagrand's *convex distance*, which can be defined for $x \in \Omega^n$ and $A \subset \Omega^n$ by

$$d_T(x, A) = \min \left\{ t : \forall \{\alpha_i\} \, \exists y \in A \text{ such that } \sum_{i=1}^n \alpha_i 1(x_i \neq y_i) \leq t \left\{ \sum_{i=1}^n \alpha_i^2 \right\}^{\frac{1}{2}} \right\}.$$

To build some intuition about this distance, we first note that if we take $\alpha_i = n^{-\frac{1}{2}}$ for all $1 \leq i \leq n$, then we find that $d_T(x, A)$ is always at least as large as $n^{-\frac{1}{2}}$ times the Hamming distance from x to A, which is given by

$$d_H(x, A) = \inf_{y \in A} \sum_{1 \leq i \leq n} 1(x_i \neq y_i).$$

One benefit of the distance $d_T(x, A)$ over $d_H(x, A)$ is that the former lets us put more weight on some of the summands than on others, but the exceptionally valuable feature of $d_T(x, A)$ is that it lets us choose weights that depend upon x. This benefit can be brought out more explicitly by writing the definition of $d_T(x, A)$ as

$$d_T(x, A) = \sup \left\{ z_\alpha : z_\alpha = \inf_{y \in A} \sum_{1 \leq i \leq n} \alpha_i(x) 1(x_i \neq y_i) \text{ and } \sum_{1 \leq i \leq n} \alpha_i^2(x) \leq 1 \right\}.$$

Shortly, we will see several examples that show how this added flexibility gives $d_T(x, A)$ a distinguished role in the probability theory of combinatorial optimization.

Talagrand (1995) developed an extensive isoperimetric theory for product spaces, and the theory provides isoperimetric inequalities for many different measures of distance, but here we will focus almost exclusively on the isoperimetric inequality for Talagrand's convex distance. In conceptual terms, the inequality says that for every $A \subset \Omega^n$ of reasonable size, the *expanded set*

$$A_t = \{x \in \Omega^n : d_T(x, A) \leq t\}$$

covers a large percentage of the space Ω^n. Since $d_T(x, A) = 0$ for $x \in A$, the essence of the matter is that $d_T(x, A)$ has sub-Gaussian tails.

THEOREM 6.1.1 (Talagrand (1995)). *For every $A \subset \Omega^n$, we have*

(6.1) $$\int_{\Omega^n} \exp\left(\frac{1}{4} d_T^2(x, A) \right) dP(x) \leq \frac{1}{P(A)},$$

and consequently,

(6.2) $$P(d_T(x, A) > t) \leq \frac{e^{-t^2/4}}{P(A)}, \quad \text{or,} \quad P(A_t) \geq 1 - \frac{e^{-t^2/4}}{P(A)}.$$

The proof of this theorem is somewhat magical, but it is not particularly difficult. Moreover, the proof follows a pattern that can be applied in many variations. In fact, this profusion of variations is one reason why Theorem 6.1.1

is best viewed as illustrative of a larger theory. In many ways, the true art of the matter rests in the framing of those isoperimetric inequalities that can serve us well in applications.

In the next section, we will use Theorem 6.1.1 to prove three strong results that make a telling case for the use of Talagrand's inequality as the tool of choice for tail bounds of combinatorial optimization. We begin by showing that Talagrand's inequality makes short work of the Gaussian tail bounds for the Steiner minimal-tree problem and for the traveling-salesman problems. These results were obtained earlier as the consequence of much more specialized arguments. Now a reasonably clever choice of the α_i is the only art that is required in order to obtain bounds that earlier required a delicate tour de force. Our third example concerns the longest-increasing-subsequence problem. This example has already helped many people understand the wisdom behind the design of $d_T(x, A)$, and in this instance the result that emerges semiautomatically via Talagrand's inequality is substantially stronger than the bounds that were obtained in Chapter 1 by Frieze's delicate use of Azuma's inequality.

After we see the ease and effectiveness with which Theorem 6.1.1 can be applied, we attend to the proof. The central idea is that natural geometric bounds on the distance $d_T(x, A)$ lead one to an induction on the dimension of the product space. The passage from one dimension to the next then boils down to a real-variable inequality which one can justify with calculus and firm resolve. After the proof, we will look at several more applications—first the theory of hereditary sets, then suprema of linear functionals of independent random variables, and finally a more sustained analysis of the nonsymmetric assignment problem.

One purely technical issue that requires mention here concerns the measurability of functions like $d_T(x, A)$ that contain suprema and infima over possibly uncountable sets. Under pathological circumstances, such functions and the sets derived from them can be nonmeasurable. Since the only product spaces that are of interest to us are either finite or equivalent to products of the unit interval, there are standard approximation techniques that can be used to avoid any measurability issues that might arise in the derivations given in this chapter. We will follow Talagrand (1995) in the custom of not commenting further on measurability.

6.2. Two geometric applications of the isoperimetric inequality.

Before we give the proof of Talagrand's inequality, we first explore some of the ways that one can design effective weight functions $\{\alpha_i\}$ for problems in combinatorial optimization. The point of this exploration is mainly to offer insight into the benefits gained by the introduction of the α_i's, but we will also make definitive progress on two problems that are of central concern in the probability theory of combinatorial optimization.

One way to think about the $\{\alpha_i\}$'s that seems to work well in many combinatorial minimization problems is that the $\alpha_i(x)$'s should measure the difficulty of dealing with the ith element of the set $x = \{x_1, x_2, \ldots, x_n\}$. We first illustrate the effectiveness of this principle with the Steiner minimal-tree problem, and then we will show that a similar approach applied to the traveling-salesman problem. The key computation is brought out quite clearly in the proof of the following general proposition.

LEMMA 6.2.1. *Suppose that L is a function on the subsets of $\mathbb{R}^d$ that is monotone in the sense that*

$$L(\{x_1, x_2, \ldots, x_n\}) \leq L(\{x_1, x_2, \ldots, x_n, x_{n+1}\}),$$

and suppose further that there are nonnegative weight functions $\alpha_i(x)$ for which for all $x = \{x_1, x_2, \ldots, x_n\}$ and $y = \{y_1, y_2, \ldots, y_n\}$, the set functional L satisfies

(6.3) $\quad L(\{x_1, x_2, \ldots, x_n\}) \leq L(\{y_1, y_2, \ldots, y_n\}) + \sum_{i=1}^{n} \alpha_i(x) 1(x_i \neq y_i).$

If the weight functions $\alpha_i(x)$ satisfy the uniform bound

(6.4) $\quad \sum_{i=1}^{n} \alpha_i(x)^2 \leq c^2,$

then for independent random variables $\{X_i : 1 \leq i \leq n\}$ that are uniformly distributed on $[0,1]^d$, we have

(6.5) $\quad P(|L(X_i, X_2, \ldots, X_n) - M_n| \geq t) \leq 4 \exp(-t^2/4c^2)$

where M_n is the median of $L(X_1, X_2, \ldots, X_n)$.

Proof. We need a set to play the role of A in Talagrand's theorem, and we can benefit from considering a natural parametrized family:

$$A(a) = \{\{y_1, y_2, \ldots, y_n\} : L(y_1, y_2, \ldots, y_n) \leq a\}.$$

By inequality (6.3), we have for all $\{x_1, x_2, \ldots, x_n\}$ and $\{y_1, y_2, \ldots, y_n\}$ that

$$L(\{x_1, x_2, \ldots, x_n\}) \leq L(\{y_1, y_2, \ldots, y_n\}) + \sum_{i=1}^{n} \alpha_i(x) 1(x_i \neq y_i),$$

so if we minimize over $y \in A(a)$, we have

$$L(x_1, x_2, \ldots, x_n) \leq a + \min_{y \in A(a)} \sum_{i=1}^{n} \alpha_i(x) 1(x_i \neq y_i)$$

$$\leq a + c d_T(x, A(a)),$$

where to derive the second inequality we first applied the Schwarz inequality with our hypothesis (6.4), and then we used the definition of $d_T(x, A(a))$. We therefore find for $x = \{X_1, X_2, \ldots, X_n\}$ that $d_T(x, A(a)) \geq c^{-1}\{L(X_1, X_2, \ldots, X_n) - a\}$.

Talagrand's inequality always gives us that

$$P(d_T(x, A(a)) \geq t) \leq \frac{1}{P(A(a))} \exp(-t^2/4);$$

we immediately find

$$P(L(X_1, X_2, \ldots, X_n) \geq a + ct) \leq \frac{1}{P(A(a))} \exp(-t^2/4).$$

In terms of $L_n = L(\{X_1, X_2, \ldots, X_n\})$, the last inequality says that

$$P(L_n \leq a)P(L_n \geq a + u) \leq \exp(-u^2/4c^2),$$

so by making the successive choices $a = M_n$ and $a = M_n - u$, the proof of the lemma is complete.

Application to the Steiner tree problem.

For our first application of Lemma 6.2.1, we consider the Steiner minimal-tree problem. In its simplest form, this problem considers a set of n points $x = \{x_1, x_2, \ldots, x_n\} \subset [0,1]^2$ and asks for the tree of minimal length that has x as a subset of its vertices. Here we might recall that the distinction between the Steiner problem and the more usual minimal-spanning-tree problem is that in the Steiner problem one gets to add any vertices that will decrease the length of the tree. We note that the length of the Steiner tree is a monotone functional in the sense of the preceding lemma. Also, for any set of points, the length of the Steiner tree is no larger than the length of the minimal spanning tree.

Now we consider the basic problem of designing a set of weight functions $\alpha_i(x)$ that will serve us well in the Steiner problem. Given $x = \{x_1, x_2, \ldots, x_n\}$, there are many natural candidates for measuring the "difficulty of serving" the point x_i, and several come to mind that directly use the geometry of the point set. For example, one might consider taking $\alpha_i(x)$ to be the length of the radius of the largest circle with center x_i which contains no other point of x.

On further reflection, there are other choices that go further in helping us make the connection to the geometry of spanning trees. Specifically, we can let $\alpha_i(x)$ denote the sum of the lengths of the edges of the minimal spanning tree of $\{x_1, x_2, \ldots, x_n\}$ that are incident to the point x_i. One of the geometrical facts that makes this choice appealing is that we know that the sum of the squares of the edges of a minimal spanning tree of n points in the unit square is bounded independently of n. We even have two proofs of this fact, one by the use of Lens geometry and one by the use of space-filling curves. Since each vertex in a minimal spanning tree in $\mathbb{R}^2$ has degree at most six, we therefore have a constant c such that for all n and all $x = \{x_1, x_2, \ldots, x_n\}$, we have

(6.6)
$$\sum_{i=1}^n \alpha_i(x)^2 \leq c^2.$$

Thus in order to prove a Gaussian tail bound for the length of the Steiner minimal tree of a random sample from $[0,1]^2$, we only need to justify the first hypothesis of Lemma 6.2.1. For a technical reason that will be evident shortly, we actually take $\alpha_i(x)$ to be *twice* the sum of the lengths of the edges that are incident to x_i in the MST of x. The only effect that this change has on inequality (6.6) is that the constant c needs to be doubled.

To prove the required bound, we first let $E_{\text{ST}(y)}$ denote the set of edges in the Steiner tree of $y = \{y_1, y_2, \ldots, y_n\}$ and let $E_{\text{MST}(x)}$ denote the set of edges of the MST of x. By our choice of $\alpha_i(x)$, the set of edges given by

$$E = E_{\text{ST}(y)} \cup \{e \in E_{\text{MST}(x)} : e = (x_i, x_j) \text{ and } x_i \notin y \text{ or } x_j \notin y\},$$

has total length that is bounded above by

$$(6.7) \qquad L(y) + \sum_{i=1}^{n} \alpha_i(x) 1(x_i \neq y_i).$$

We first note that if $x \cap y = \emptyset$, then inequality (6.3) certainly holds since the right-hand side is at least twice the length of the MST of x. Next, since L is monotone, we have $L(x) \leq L(x \cup y)$, so to prove that L satisfies the hypothesis (6.3), we just need to show that for $x \cap y \neq \emptyset$ and $x \neq y$ the graph G with vertex set $x \cup y$ and edge set E is connected.

We first note that the set y is contained in a single connected component of G since the edge set of G contains $E_{\text{ST}(y)}$. Since $x \cap y \neq \emptyset$, we can choose an $x_0 \in x$ with $x_0 \in y$, and for any other $x_i \in x$ we can consider the path P in the MST of x that goes from x_i to x_0. If $e = (x_i, x_j)$ is the first edge of this path, then e is in E since one of its endpoints is outside of y and since $e \in \text{MST}(x)$. Consider the next edge e' of the path. Either it connects e to a point of y or else $e' \in E$ since e' is an edge of the MST(x) for which one end is not in y. We can continue along the path P proving that each successive edge is either in E or else meets y. Since the first edge to meet y also has one of its vertices outside of y, we therefore conclude that P contains (as an initial segment) a path of edges in E that goes from x_0 to y. This proves that all the vertices of $x \cup y$ are in the same connected component of G as y. Thus G is connected. We have therefore established that the Gaussian tail bound (6.5) holds for the Steiner tree problem for the unit square in $\mathbb{R}^2$.

Application to the TSP.

Tail bounds for the TSP have provided a basic testbed for many of the tools that are used in the probability theory of combinatorial optimization. Though such bounds were not directly pursued in Beardwood, Halton, and Hammersley (1959), they were naturally implicit. The jackknife inequality of Efron and Stein (1981) led to the observation in Steele (1981b) that for random samples from $[0, 1]^2$, the variance of the tour length is bounded independently of n. Martingales and interpolation theory were then used in Rhee and Talagrand (1987, 1989a) to get the first concentration inequalities that were sharp enough to credibly suggest that the TSP might have Gaussian tails. Finally, in Rhee and Talagrand (1989b), the basic elements of the martingale approach were combined with an intricate geometric argument to give the first proof that the length of the TSP tour for random samples from $[0, 1]^2$ has a Gaussian tail bound.

Now by using Talagrand's convex-distance inequality as repackaged in inequality (6.5) of Lemma 6.2.1, one can now give a remarkably easy proof of the Gaussian tail bound for the TSP. Within the framework of the convex-distance inequality, the key point boils down to making a sensible choice for the $\alpha_i(x)$. As in the Steiner tree problem, the most effective choice seems to be one that gets at the notion of the "difficulty to serve" the point x_i, but again we want a measure that relates as directly as possible to the specific geometry of the problem. Such considerations still provide many alternatives, but again one that serves readily and well is provided by the space-filling curve heuristic. Specifically, given $x = \{x_1, x_2, \ldots, x_n\} \subset [0, 1]^2$, we let $\alpha_i(x)$ denote twice the length of the two edges that are incident to x_i in the tour of the set x that is given by the space-filling curve heuristic. In parallel to our analysis of the Steiner tree problem, we let $L(x)$ denote the length of the shortest tour through the set x, and we note that for any finite $y \subset [0,1]^2$ we have the bound $L(x) \leq L(x \cup y)$. We already know from our analysis of the space-filling curve heuristic that the Lipschitz property of the space-filling curve mapping guarantees that for all $x = \{x_1, x_2, \ldots, x_n\} \subset [0, 1]^2$, the functions $\alpha_i(x)$ satisfy the uniform bound

$$\sum_{i=1}^{n} \alpha_i(x)^2 \leq c^2.$$

To complete our justification that our choice of $\alpha_i(x)$ will give us a Gaussian tail bound, all that remains is to show that we have

(6.8) $\quad L(\{x_1, x_2, \ldots, x_n\}) \leq L(\{y_1, y_2, \ldots, y_n\}) + \sum_{i=1}^{n} \alpha_i(x) 1(x_i \neq y_i).$

An easy way to see why inequality (6.8) holds is to consider the schematic layout given in Figure 6.1. An ellipse is used to indicate the space-filling curve tour of the set x, and each of the $x_i \in x$ is denoted by a black dot on the ellipse. The points on the ellipse that are covered with small boxes are the elements of $x \cap y$, and we can assume that $x \cap y \neq \emptyset$, or else with our choice of $\alpha_i(x)$, inequality (6.8) would be trivial. Now from each box in the figure (or each element of $x \cap y$), we consider the circular path that is indicated by the pairs of arrows along the ellipse. Each of the sets of arrow edges forms a cycle and the union U of these cycles covers the set of vertices in $x \cap y^c$. We also note that each of the cycles contains exactly one element of y.

If we consider the optimal tour T of y, then the graph with edge set $T \cup U$ and vertex set $x \cup y$ is therefore a connected graph with even degree for every vertex. Such a graph contains a tour T' of the vertex set $x \cup y$, and the length of T' is certainly an upper bound on the length of the optimal tour through the points of x. Since the sum of the the edges in the set $T \cup U$ is bounded above by the right-hand side of inequality (6.8), the proof of inequality (6.8) is complete.

6.3. Application to the longest-increasing-subsequence problem.

In our justification of the Gaussian tail bound for the TSP and Steiner problems, our design of the weight functions was guided by geometry, but in the analysis of the longest-increasing-subsequence problem, the weight functions are given by combinatorial considerations. It turns out that one does rather well with a straightforward choice of the weights that take on only two values.

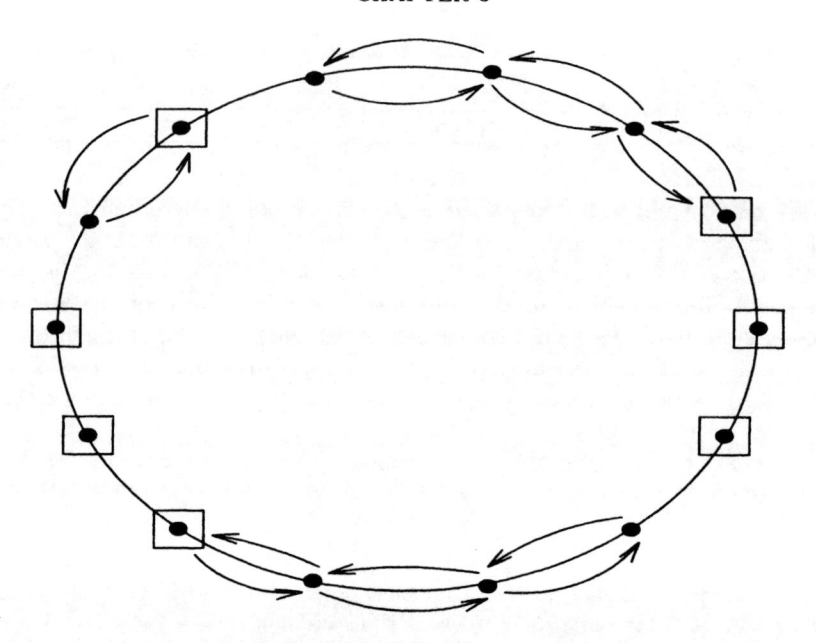

FIG. 6.1. *Using space-filling curves to make a distance.*

Just as in Chapter 1, we let $I_n(x)$ denote the cardinality of the longest increasing subsequence of $x = (x_1, x_2, \ldots, x_n)$, so if $I_n(x) = k$ there is a set $J = \{i_1, i_2, \ldots, i_k\} \subset [n]$ such that $x_{i_1} < x_{i_2} < \cdots < x_{i_k}$. Because of its ability to focus on J, Talagrand's isoperimetric inequality will have an easy time improving on our earlier analysis of the tail behavior of $I_n(x)$.

For any y, the length of the longest increasing subsequence for y is at least as large as the difference between $I_n(x) = k$ and the number of places in J where x and y differ, so we have

$$I_n(y) \geq I_n(x) - \sum_{i \in J} 1(x_i \neq y_i).$$

The last sum can be bounded in terms of $d_T(x, A(a))$ for $A(a) = \{y : I_n(y) \leq a\}$ since if we let $\alpha_i = 1/\sqrt{k}$ for $i \in J$ and otherwise take $\alpha_i = 0$, we find from the definition of d_T that

$$\min_{y \in A(a)} \sum_{i \in J} 1(x_i \neq y_i) = \sqrt{k} \min_{y \in A(a)} \sum_{i \in J} \frac{1}{\sqrt{k}} 1(x_i \neq y_i) \leq d_T(x, A(a))\sqrt{k}.$$

Therefore, we see that for any x there is a $y^* \in A(a) = \{y : I_n(y) \leq a\}$ such that

$$a \geq I_n(y^*) \geq I_n(x) - d_T(x, A(a))\sqrt{I_n(x)},$$

and this last expression is tidier if rewritten without intermediation of y^* to give an inequality that is valid for all $a \geq 0$ and x:

(6.9) $$d_T(x, A(a)) \geq \frac{I_n(x) - a}{\sqrt{I_n(x)}}.$$

The function $g_a(u) = (u-a)/\sqrt{u}$ is monotone increasing for $u \geq a$, so for any $a \geq 0$ we have

$$\begin{aligned} P(I_n(x) \geq a + u) &= P\left(\frac{I_n(x) - a}{\sqrt{a+u}} \geq \frac{u}{\sqrt{a+u}}\right) \\ &\leq P\left(d_T(x, A(a)) \geq \frac{u}{\sqrt{a+u}}\right) \\ &\leq \frac{1}{P(A(a))} \exp\left(-\frac{u^2}{4(a+u)}\right), \end{aligned}$$

where in the second step we used (6.9) and in the last step we applied Talagrand's inequality (6.2). If M is any median (so that M satisfies $P(I_n \leq M) \geq \frac{1}{2}$ and $P(I_n \geq M) \geq \frac{1}{2}$), then the last inequality can be written as

(6.10) $$P(I_n(x) \geq M + u) \leq 2\exp\left(-\frac{u^2}{4(M+u)}\right).$$

To work toward an inequality that goes in the other direction, we note that by monotonicity and (6.9) we have

$$\begin{aligned} I_n \geq M &\iff \frac{I_n(x) - a}{\sqrt{I_n(x)}} \geq \frac{M - a}{\sqrt{M}} \\ &\iff \frac{I_n(x) - M + u}{\sqrt{I_n(x)}} \geq \frac{u}{\sqrt{M}} \\ &\implies d_T(x, \{I_n \leq M - u\}) \geq \frac{u}{\sqrt{M}}. \end{aligned}$$

This chain of inferences and Talagrand's inequality tell us that

$$\frac{1}{2} \leq P\left(d_T(x, \{I_n \leq M - u\}) \geq \frac{u}{\sqrt{M}}\right) \leq \frac{\exp(-u^2/4M)}{P(I_n \leq M - u)}$$

or, in other words, we have

(6.11) $$P(I_n \leq M - u) \leq 2\exp\left(\frac{-u^2}{4M}\right).$$

Inequalities (6.10) and (6.11) show that I_n is sharply concentrated in an interval about M of size $O(M^{\frac{1}{2}}) = O(n^{\frac{1}{4}})$. This is a stronger result than the bound that we obtained in Chapter 1 with much more work using the "flipping" method based on Azuma's inequality.

The argument just given for I_n does not use much of the structure of the longest-increasing-subsequence problem; and, as Talagrand (1995) points out, there is a broad class of natural random variables to which the preceding argument may be applied without any change. If a map L_n from Ω^n to the set of integers has the property that for every $x = (x_1, x_2, \ldots, x_n)$ there is a $J \subset [n]$ with card $J = L_n(x)$ such that for any $y \in \Omega^n$ we have

$$L_n(y) \geq \text{card}\{i \in [n] : y_i = x_i\},$$

we will call L_n a *configuration function*. Reconsideration of the proof of the tail bounds for I_n will show that (6.9), (6.10), and (6.11) are all valid for any configuration function. For example, these inequalities can be applied to the length of the longest *unimodal subsequence* of a sequence of independent uniformly distributed random variables.

6.4. Proof of the isoperimetric inequality.

Before we begin the proof of Talagrand's theorem, we need to develop a richer view of the distance $d_T(x, A)$. In particular, we will benefit from a dual characterization of $d_T(x, A)$ as the distance from the origin to a convex set that is determined by A.

Perhaps the most natural candidate for a set in $\mathbb{R}^n$ that can capture the relationship between x and A is the set of all of the 0-1 vectors $v_y = (1(x_1 \neq y_1), 1(x_2 \neq y_2), \ldots, 1(x_n \neq y_n))$ with $y \in A$. One can preserve the intuition behind this candidate and gain some technical advantages by considering a larger set. The idea is to include all of the vectors $v_y = (1(x_1 \neq y_1), 1(x_2 \neq y_2), \ldots, 1(x_n \neq y_n))$ with $y \in A$ together with the vectors that can be formed from these by conversion of some of the 0's to 1's. Formally, we let $U_A(x)$ denote the set of 0-1 vectors of length n such that $u \in U_A(x)$ if and only if there is a $y \in A$ such that u has 1's in at least all of the places that we have 1's in some $v_y = (1(x_1 \neq y_1), 1(x_2 \neq y_2), \ldots, 1(x_n \neq y_n))$. In other words, $v \in U_A(X)$ if and only if $v - v_y \geq 0$ for some $y \in A$. In language that is often used in combinatorics, $U_A(x)$ is the set of all 0-1 vectors that dominate $v_y = (1(x_1 \neq y_1), 1(x_2 \neq y_2), \ldots, 1(x_n \neq y_n))$ for some $y \in A$. Finally, we let $V_A(x)$ denote the convex hull of $U_A(x)$.

PROPOSITION 6.4.1 (dual characterization of $d_T(x, A)$).

$$d_T(x, A) = \min\{\|v\|_2 : v \in V_A(x)\}.$$

Proof. Suppose we are given a vector α with $\alpha_i \geq 0$ for all $1 \leq i \leq n$. If we first use the definition of $U_A(x)$, then use the fact that the minimum of a linear functional on a convex set equals the minimum over the set of extreme points, and finally apply Schwarz's inequality, we find

$$\min_{y \in A} \sum_{i=1}^n \alpha_i 1(x_i \neq y_i) = \min_{u \in U_A(x)} \sum_{i=1}^n \alpha_i u_i$$

$$= \min_{v \in V_A(x)} \sum_{i=1}^n \alpha_i v_i$$

$$\leq \min_{v \in V_A(x)} \left\{\sum_{i=1}^n \alpha_i^2\right\}^{1/2} \left\{\sum_{i=1}^n v_i^2\right\}^{1/2}$$

$$= \left\{\sum_{i=1}^n \alpha_i^2\right\}^{1/2} \min\{\|v\|_2 : v \in V_A(x)\}.$$

The last inequality tells us that $d_T(x, A) \leq \min\{\|v\|_2 : v \in V_A(x)\}$ by the definition of $d_T(x, A)$.

To prove the reverse inequality, we let $\tau = \min\{\|v\|_2 : v \in V_A(x)\}$ and note by the linear functional characterization of the Euclidean norm (or by the Hahn–Banach theorem), there is an α with $\{\sum_{i=1}^n \alpha_i^2\}^{\frac{1}{2}} = 1$ such that for all $v \in V_A(x)$, we have

$$\sum_{i=1}^n \alpha_i v_i \geq \tau,$$

so in particular we have for all $y \in A$ that

$$\sum_{i=1}^n \alpha_i 1(x_i \neq y_i) \geq \tau.$$

The last expression tells us exactly what is required to say that $d_T(x, A) \geq \tau$, so the proof of the proposition is complete.

We now take on the proof of Theorem 6.1.1. As promised, we will use induction, so we may as well first check the validity of the theorem in the base case $n = 1$. For $n = 1$, we have

$$d_T(x, A) = \begin{cases} 0 & \text{for } x \in A, \\ 1 & \text{for } x \notin A, \end{cases}$$

so the integral of Talagrand's theorem becomes simply

$$\int \exp\left(\frac{1}{4} d_T^2(x, A)\right) dP(x) = e^{1/4}(1 - P(A)) + P(A).$$

Now for all $0 \leq u \leq 1$, we have $(1-u)e^{\frac{1}{4}} + u \leq u^{-1}$ by easy calculus, so the proof of the case $n = 1$ is complete.

We take the assertion of Theorem 6.1.1 for n as our induction hypothesis, and we work toward the proof of the inequality for $n + 1$. We next take $A \subset \Omega^{n+1}$, and for a typical point of $\Omega^{n+1} = \Omega^n \times \Omega$, we write (x, ω) with $x \in \Omega^n$ and $\omega \in \Omega$. There are two sets that are central to the understanding of the geometry of product spaces: *the ω section of A*,

$$A(\omega) \equiv \{x : (x, \omega) \in A\} \subset \Omega^n,$$

and *the projection of A*,

$$B \equiv \bigcup_{\omega \in \Omega} A(\omega) = \{x : \exists \, (x, \omega) \in A\} \subset \Omega^n.$$

At the heart of the proof of Talagrand's theorem is the fact that $d_T(x, A)$ can be bounded in terms of corresponding convex distances for the sections and the projection. Specifically, we will show that for all $z = (x, \omega) \in \Omega^{n+1}$ and all $0 \leq \lambda \leq 1$, we have the key inequality

(6.12) $\quad d_T^2(z, A) \leq (1 - \lambda) d_T^2(x, B) + \lambda d_T^2(x, A(\omega)) + (1 - \lambda)^2.$

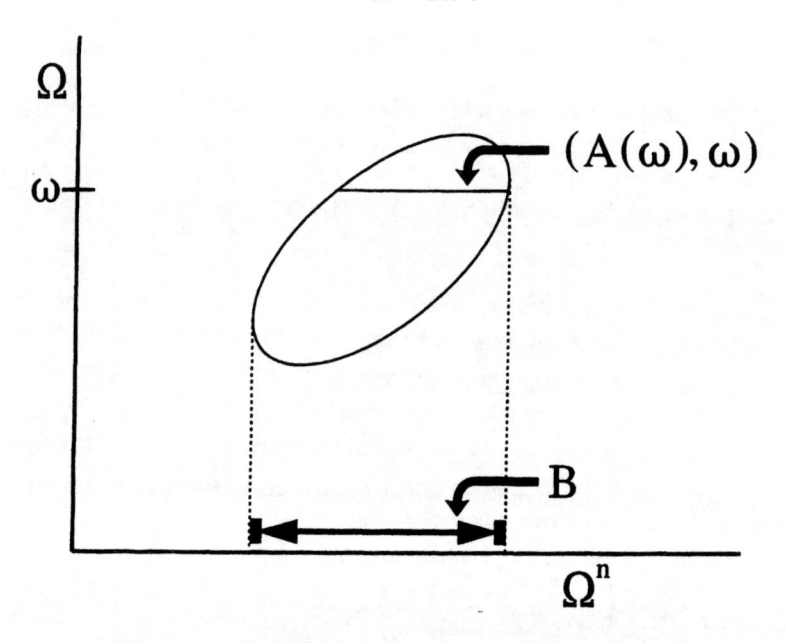

FIG. 6.2. *Sections and projections used in the proof.*

To prove (6.12) we first use the dual characterization of d_T to permit us to choose two vectors $v' \in V_{A(\omega)}(x)$ and $v'' \in V_B(x)$ such that

(6.13) $\qquad \|v'\| = d_T(x, A(\omega)) \quad \text{and} \quad \|v''\| = d_T(x, B).$

Now for any $s \in U_{A(\omega)}(x)$ we have $(s, 0) \in U_A(z)$ and for any $t \in U_B(x)$ we have $(t, 1) \in U_A(z)$; hence for any $s \in V_{A(\omega)}(x)$ and $t \in V_B(x)$, we see from the definition of the V-sets that $(s, 0)$ and $(t, 1)$ are both in $V_A(z)$. Since $V_A(z)$ is convex, we have for any $0 \leq \lambda \leq 1$ that

$$\lambda(s, 0) + (1 - \lambda)(t, 1) \in V_A(z).$$

When we apply this fact to the $v' \in V_{A(\omega)}(x)$ and $v'' \in V_B(x)$ of (6.13), we find

$$v''' \equiv \lambda(v', 0) + (1 - \lambda)(v'', 1)) \in V_A(z).$$

By the duality characterization of d_T, the length of v''' is an upper bound on $d_T(z, A)$, so

$$d_T^2(z, A) \leq \|v'''\|_2^2 = \sum_{i=1}^{n} (\lambda v'_i + (1 - \lambda) v''_i)^2 + (1 - \lambda)^2.$$

Now since $(\lambda a + (1 - \lambda) b)^2 \leq \lambda a^2 + (1 - \lambda) b^2$, we see that

$$d_T^2(z, A) \leq \lambda \|v'\|_2^2 + (1 - \lambda) \|v'\|_2^2 + (1 - \lambda)^2,$$

and since by (6.13) the norms of v' and v'' equal $d_T(x, A(\omega))$ and $d_T(x, B)$, the proof of the key inequality (6.12) is complete.

Now we are prepared for the main induction step. We keep ω fixed and note by Hölder's inequality that the n-fold integral

$$I_n(\omega) = \int_{\Omega^n} \exp\frac{1}{4} d_T((x,\omega), A) \, dP(x)$$

is bounded by

$$\exp\frac{1}{4}(1-\lambda)^2 \left(\int_{\Omega^n} \exp\frac{1}{4} d_T(x, A(\omega)) \, dP(x)\right)^\lambda \left(\int_{\Omega^n} \exp\frac{1}{4} d_T(x, B) \, dP(x)\right)^{1-\lambda}.$$

We can now apply the induction hypothesis to both of the last two integrals to find that for $I_n(\omega)$ we have the pointwise upper bound

$$\exp\frac{1}{4}(1-\lambda)^2 \left(\frac{1}{P(A(\omega))}\right)^\lambda \left(\frac{1}{P(B)}\right)^{1-\lambda} = \frac{1}{P(B)} \left(\frac{P(A(\omega))}{P(B)}\right)^{-\lambda} \exp\frac{1}{4}(1-\lambda)^2.$$

The last few steps have been natural enough, but now we need to think. To complete the induction, we will have to integrate over ω, and the only exact integration that seems easy is the integration of $P(A(\omega))$ to get $P(A)$. This suggests that when we optimize over λ, the critical issue is to get an expression that can be bounded by a linear function of $P(A(\omega))$. Exploration is needed to decide upon a proper bound, but in finite time one is led to the following lemma.

LEMMA 6.4.1. *For all $0 \leq r \leq 1$, we have*

$$\inf_{0 \leq \lambda \leq 1} r^{-\lambda} \exp\frac{1}{4}(1-\lambda)^2 \leq 2 - r.$$

Calculus and care suffice to prove the lemma, so we will skip its proof. Instead, we apply the lemma to the pointwise upper bound that we found for $I_n(\omega)$. We find that

$$I_n(\omega) = \int_{\Omega^n} \exp\frac{1}{4} d_T((x,\omega), A) \, dP(x) \leq \frac{1}{P(B)} \left(2 - \frac{P(A(\omega))}{P(B)}\right),$$

which according to plan we can easily integrate with respect to ω to find

$$\int_{\Omega^{n+1}} \exp\frac{1}{4} d_T((x,\omega), A) \, dP(x) dP(\omega) \leq \frac{1}{P(B)} \left(2 - \frac{P(A)}{P(B)}\right).$$

Finally, if we multiply and divide the last expression by $P(A)$, we get an upper bound that can be written as $P(A)^{-1} f(P(A)/P(B))$, where $f(x) = x(2-x)$. Since $f(x) = x(2-x) \leq 1$ for all real x, the proof of the induction step is complete.

6.5. Application and comparison in the theory of hereditary sets.

If y and z are elements of $\{0,1\}^n$, we say that z is a *descendant* of y and write $z \prec y$ if for all $1 \leq i \leq n$ we have $z_i \leq y_i$. A set $A \subset \{0,1\}^n$ is called a *hereditary set* if $y \in A$ and $z \prec y \Rightarrow z \in A$. Consistent with most Western hereditary traditions, if you are a descendant of a Smith, then you are a Smith. Hereditary sets often arise in combinatorial investigations, and examples abound. To give one, we note that the set of all $y \in \{0,1\}^n$ such that y has no more than three consecutive 1's is a hereditary set.

For hereditary sets, the Hausdorff distance has a particularly nice relationship to Talagrand's convex distance. As a consequence, the isoperimetric inequality provides a most informative tool for understanding probabilities associated with hereditary sets. The observation that gets the ball rolling is that for a hereditary A, we have

$$(6.14) \quad d_H(x,A) = \min_{y \in A} \sum_{i=1}^n 1(y_i \neq x_i) = \min_{y \in A} \sum_{i=1}^n 1(y_i \neq x_i)1(x_i = 1)$$

because for any y any disagreement between x_i and y_i with $x_i = 0$ could be resolved by choosing an element of $y' \in A$ that agrees with y except for changing y_i to 0. Now for any weights $\{\alpha_i(x_i)\}$ with $\sum_{i=1}^n \alpha_i(x_i)^2 \leq 1$, we have by the definition of $d_T(x,A)$ that

$$d_T(x,A) \geq \min_{y \in A} \sum_{i=1}^n \alpha_i(x_i) 1(y_i \neq x_i),$$

so if we take $\alpha_i(x_i) = 1(x_i = 1)\left(\sum_{i=1}^n 1(x_i = 1)\right)^{-\frac{1}{2}}$, we find with help from (6.14) that

$$d_T(x,A)\left(\sum_{i=1}^n 1(x_i = 1)\right)^{1/2} \geq \min_{y \in A} \sum_{i=1}^n 1(y_i \neq x_i)1(x_i = 1) = d_H(x,A).$$

We therefore find for any hereditary A and all k, we have

$$P(d_H(x,A) \geq t) \leq P(d_T(x,A) \geq t/\sqrt{k}) + P\left(\sum_{i=1}^n 1(x_i = 1) \geq k\right),$$

and by Talagrand's inequality we therefore find that for all $t \geq 0$ and all $k \geq 0$, we have

$$(6.15) \quad P(d_H(x,A) \geq t) \leq \frac{1}{P(A)} e^{-t^2/4k} + P\left(\sum_{i=1}^n 1(x_i = 1) \geq k\right).$$

Some comparisons.

The general inequality (6.15) should now be compared with results that have been developed in the theory of hereditary sets. The comparison is a natural one since some of the first isoperimetric inequalities of combinatorial theory we developed in the context of hereditary sets. Leader (1991) provides an excellent survey of that development, so we can focus on just one of the more refined results due to Bollobás and Leader (1991a, b).

THEOREM 6.5.1 (Bollobás and Leader (1991a)). *Let P denote the measure on $\{0,1\}^n$ given by the n-fold product of μ, where $\mu\{1\} = p$ and $\mu\{0\} = q$ with $p + q = 1$ and $0 < p < 1$. If A is any hereditary set with $P(A) \geq \frac{1}{2}$, then*

$$(6.16) \quad P(d_H(x,A) \geq t) \leq \frac{t}{(pqn)^{1/2}} e^{-t^2/2pqn}$$

for
$$(pqn)^{1/2} \leq t \leq \min\left(\frac{pqn}{10}, \frac{1}{2}(pqn)^{2/3}\right).$$

The precision of Theorem 6.5.1 is easiest to guage in comparison to high-precision estimates for the binomial tails. Since the sets $\{x = (x_1, x_2, \ldots, x_n) : \sum x_i \leq a\}$ are hereditary, Theorem 6.5.1 gives us a bound on the binomial tail probabilities, and we can check that the bound we obtain compares well with the best. Specifically, if $S_{n,p} = \sum_{i=1}^{n} \xi_i$, where the ξ_i's are independent with $P(\xi_i = 1) = p = 1 - P(\xi_i = 0)$, then a sharp form of the traditional tail bound for the binomial distribution gives us

$$(6.17) \quad P(S_{n,p} \geq pn + t) \leq \left(\frac{pqn}{2\pi}\right)^{1/2} \frac{1}{t} \exp\left(\frac{-t^2}{2pqn} + \frac{t}{pqn} + \frac{t^3}{p^2 n^2}\right),$$

provided that $1 \leq pn$ and $1 \leq t \leq qn/3$ (cf. Bollobás (1985, p. 10)).

In both (6.16) and (6.17), we see that for small p an essential feature of the inequalities is that the coeficient of t^2 is of order $1/pn$. Naturally, the delicate quality of (6.16) permits even more precise comparison, but the key qualitative aspect of the two bounds is reflected the order of the coefficient of t^2. In the corrolary (6.15) of our general isoperimetic inquality, we find the same fundamental behavior; specifically, if we take $k = pn + t$ in (6.15), we find from (6.17) that the dominant term of (6.15) is the first term, and for $0 < t \leq 4np$ that term is further dominated by $P(A)^{-1} \exp(-t^2/8pn)$.

The punchline of this anaylsis is that we have bartered very little precision in exchange for the generality we gain in Theorem 6.1.1. In the theory of hereditary sets, we can recover the essential behavior of delecate results that were developed using specialized tools. Moreover, we will find that Theorem 6.1.1 yields near optimal results in more examples than we have any right to expect.

6.6. Suprema of linear functionals.

We will next explore the application of Talagrand's isoperimetric inequality to a class of problems that directly involve sums of independent random variables. At first blush, such problems may seem too classical to fit well with the theory that is developed here, but very shortly we will find that there are some direct correspondances. Let $\mathcal{F}$ denote a family of n-tuples $\alpha = (\alpha_1, \alpha_2, \ldots, \alpha_n)$ and let $\tau = \sup_{\alpha \in \mathcal{F}} (\sum_{i=1}^{n} \alpha_i^2)^{\frac{1}{2}}$. We will be especially interested in the sums of the form

$$(6.18) \quad Z = \sup_{\alpha \in \mathcal{F}} \sum_{i=1}^{n} \alpha_i X_i.$$

With just a small amount of translation, one can easily see the relevance of sums like (6.18) to many problems like those that have been studied in earlier chapters. To spell out this connection, we first take the complete graph on $\{1, 2, \ldots, n\}$ and let c_{ij} be the cost associated with edge (i, j). For specificity, we may as well consider the n^2 variables c_{ij} to be independent and uniformly distributed on $[0, 1]$. Now let $\mathcal{G}$ be any class of subgraphs of the complete graph and consider the variable defined by

$$Z_\mathcal{G} = \inf_{G \in \mathcal{G}} \sum_{(i,j) \in E(G)} c_{ij}.$$

This variable is of precisely the kind addressed (6.18), as one can see by associating α with G by taking α to be the indicator vector for the set of edges of G and by the judicious use of minus signs to convert the supremum into an infimum.

THEOREM 6.6.1. *Let X_i, $1 \leq i \leq n$, be independent random variables that satisfy the bounds $r_i \leq X_i \leq r_i + 1$. We also let*

$$Z = \sup_{\alpha \in \mathcal{F}} \sum_{i=1}^n \alpha_i X_i,$$

where the set $\mathcal{F}$ of real weights $\alpha = (\alpha_1, \alpha_2, \ldots, \alpha_n)$ is only required to satisfy

$$\tau \equiv \sup_{\alpha \in \mathcal{F}} \left(\sum_{i=1}^n \alpha_i^2 \right)^{1/2} < \infty.$$

For all $a \geq 0$ and $b \geq 0$, we have the symmetrical tail bound

(6.19) $$P(Z \geq b) P(Z \leq a) \leq \exp\left(-\frac{(b-a)^2}{4\tau^2} \right).$$

Before going into the proof, we should note that tail bounds like (6.19) appear quite naturally when one uses Talagrand's isoperimetric inequality, even though such bounds may look strange from a traditional point of view. Still, we can easily convert the tail bound (6.19) into one of the traditional kind. Specifically, if M is any median of Z, we just consider the two inequalities that result from applying (6.19) with the assignment sets $a = M, b = u$ and $a = u, b = M$. We then find that for all $u \geq 0$, we have

(6.20) $$P(|Z - M| \geq u) \leq 2 \exp\left(-\frac{u^2}{4\tau^2} \right).$$

To prove Theorem 6.6.1, we first take any $x = (x_1, x_2, \ldots, x_n)$ and then define the function $Z(x) = \sup_{\alpha \in \mathcal{F}} \sum_{i=1}^n \alpha_i x_i$. We then let $A(a) = \{y : Z(y) \leq a\}$ and note that for any α, either in $\mathcal{F}$ or not, the definition of $d_T(x, A(a))$ tells us that

$$d_T(x, A(a)) \geq \min_{y \in A(a)} \sum_{i=1}^n \left\{ |\alpha_i| \left(\sum_{i=1}^n \alpha_i^2 \right)^{-1/2} \right\} 1(x_i \neq y_i).$$

For some $y \in A(a)$, we will then certainly have

$$\sum_{i=1}^n |\alpha_i| 1(x_i \neq y_i) \leq \tau d_T(x, A(a)).$$

But since x_i and y_i are both in $[0,1]$, we have the crude bound $|x_i - y_i| \leq 1(x_i \neq y_i)$, from which we see

$$\left| \sum_{i=1}^n \alpha_i(r_i + y_i) - \sum_{i=1}^n \alpha_i(r_i + x_i) \right| \leq \sum_{i=1}^n |\alpha_i| \, |x_i - y_i| \leq \tau d_T(x, A(a)).$$

The last inequality implies that

$$\sum_{i=1}^n \alpha_i(r_i + x_i) \leq Z(y) + \tau d_T(x, A(a)),$$

so Talagrand's Gaussian tail bound on $d_T(x, A(a))$ tell us precisely that

$$P(Z(x) \geq b) P(Z(x) \leq a) \leq \exp\left(-\frac{(b-a)^2}{4\tau^2}\right),$$

completing the proof of Theorem 6.6.1.

Inequality (6.20) carries a great deal of useful information, but sometimes one has to work a bit before its implications can be seen in the best light. For example, with a typical interesting choice of $\mathcal{G}$, the naïve application of Theorem 6.6.1 is likely to be disappointing. Thus if we let $\mathcal{G}$ be the set of spanning trees of the complete graph on n vertices, a moment's reflection will show that the direct application of Theorem 6.6.1 is too crude. For this example and many other interesting choices, the expectation of the variable $Z_\mathcal{G}$ is bounded independently of n, but the naïve application of Theorem 6.6.1 would tell us only that $Z_\mathcal{G}$ is concentrated in an interval of size $O(\sqrt{n})$.

The key to extracting more meaningful bounds from Theorem 6.6.1 in such problems is to take advantage of the fact that the edges with cost c_{ij} larger than a certain threshold are unlikely to be used. In the next section, we pursue this suggestion for the assignment problem, and we will find that Theorem 6.6.1 does indeed lead to useful tail bounds. In fact, Theorem 6.6.1 yields the best bounds that are known.

6.7. Tail of the assignment problem.

The most basic random variable associated with the assignment problem is the bottom-line cost

$$A_n = \min_\sigma \sum_{i=1}^n c_{i\sigma(i)},$$

where the minimum is over all permutations of $[n] = \{1, 2, \ldots, n\}$ and the c_{ij}'s are n^2 independent random variables with the uniform distribution on $[0,1]$. We have seen several proofs of the fact that EA_n is bounded, and in particular we know from Karp's proof that $EA_n < 2$. Also, in our investigation of the greedy algorithms for the assignment problem, we found that in any assignment one is likely to need some edges with cost at least as big as $\log n / n$. Still, the same calculations gave a strong suggestion that one should often be able to find an assignment that does not use any edges that are appreciably larger than $(\log n)^\beta/n$ for some $\beta \geq 1$.

These considerations lead us to introduce some new cost variables by taking

$$c'_{ij} = \min(c_{ij}, p),$$

where p is a quantity that goes to zero with n. We will find shortly that with very high probability we get the same cost for the minimal assignment by using the costs c'_{ij} that we get when we use costs c_{ij} provided that we take p to be of the order of $(\log n)^2/n$.

The proof of this fact seems to be more involved than one might guess from our analysis of the greedy algorithms, but in the course of the proof we will develop some general tools that are useful in many related investigations. The first result that we need boils down to the the fact that in random bipartite graphs where the probability of an edge is p, we are very likely to have lots of of reasonably short cycles.

The formal development of this idea requires some notation. Consider a bipartite graph G with bipartition (A, B), where $\operatorname{card} A = \operatorname{card} B = n$. For any set $S \subset A$, we let $\Gamma(S)$ denote the set of neighbors of the set S; that is, we have $\Gamma(S) = \{v \in B : (u, v) \in E(G) \text{ and } u \in S\}$. We will call the bipartite graph G an α-expander if we have for all $S \subset A$ that

(6.21) $\qquad \operatorname{card} S \leq n/2 \Rightarrow \operatorname{card} \Gamma(S) \geq \min(n/2, \alpha \operatorname{card} S)$

and

(6.22) $\qquad \operatorname{card} S \geq n/2 \Rightarrow (n - \operatorname{card} \Gamma(S)) \leq \frac{1}{\alpha}(n - \operatorname{card} S).$

Expander graphs have many good properties, and in particular the next lemma shows that for any matching M of the complete graph on (A, B), the graph $M \cup G$ can be covered by a disjoint collection of small cycles. The lemma is phrased in a way that anticipates its application to the assignment problem.

LEMMA 6.7.1 (alternating-chain lemma). *Suppose that the bipartite graph G has bipartition (A, B) with $\operatorname{card} A = \operatorname{card} B = n$. Suppose that a one-to-one mapping $\tau : A \to B$ is given. If G is an α-expander and k is the least integer such that $\alpha^k > n/2$, then given any $i \in A$ there is an $m \leq 2k$ and there are m distinct points $\{i_s \in A : 1 \leq s \leq m\}$ such that if we set $i_{m+1} = i$ then we have $i_1 = i$ and $(i_s, \tau(i_{s+1})) \in E(G)$ for all $1 \leq s \leq m$.*

Proof. For any $i \in A$, we let $S_p(i)$ denote the set of all $t \in A$ such that there exists a sequence of points $i_1, i_2, \ldots, i_p \in A$ that begins at i (so $i_1 = i$), ends at t (so $i_p = t$), and has the property that $(i_s, \tau(i_{s+1}))$ is an edge of G for all $1 \leq s < p$. By the definition of $S_{p+1}(i)$, we see that $\tau^{-1}\{\Gamma(S_p(i))\} \subset S_{p+1}(i)$; so since τ is a one-to-one mapping, we see that

(6.23) $\qquad \operatorname{card} \Gamma(S_p(i)) \leq \operatorname{card} S_{p+1}(i).$

By the definition of k, we have $\alpha^{k-1} \leq n/2$, so by iteration of (6.21) and by (6.23), we have $\operatorname{card} S_p(i) \geq \alpha^{p-1}$ for all $p < k$. Hence we have $\operatorname{card} S_{k+1}(i) \geq n/2$. We then use (6.22) to note for all $p \geq 1$ that

$$n - \operatorname{card} S_{k+1+p}(i) \leq \alpha^{-p} n/2,$$

and since $\alpha^{-k}n/2 < 1$, we see that card $S_{2k+1}(i) = n$. This tells us in particular that $i \in S_{2k+1}(i)$, so if we if we let m be the smallest integer such that $i \in S_{m+1}(i)$, we have $m \leq 2k$. By the minimality of m, we have that if we write $S_{m+1}(i) = \{i_1, i_2, \ldots, i_{m+1}\}$ we have that all of the i_j's are distinct for $1 \leq j \leq m$ and that $i = i_{m+1}$; moreover, by the definition of $S_p(i)$, we have $\imath = i_1$ and $(i_s, \tau(i_{s+1}))$ for all $1 \leq s \leq m$, so the lemma is proved.

The next proposition tells us how to use of information on the expansion property for the subgraph with short edges in order to get a bound on the size of the largest edge in the optimal assigment. The bound given by the lemma will then let us use our results on suprema of linear funtionals to obtain a good concentration properties for A_n.

PROPOSITION 6.7.1. *Let G denote a weighted bipartite graph with nonnegative edge weights w_{ij} and bipartition (A, B) where* card $A =$ card $B = n$, *and let G' denote the bipartite graph with the same bipartition (A, B) but with the edge set given by*

$$E(G') = \{(i, j) : w_{ij} \leq \lambda\}.$$

If G' is an α-expander graph, then for any optimal assignment τ of G, we have

(6.24) $$w_{i\tau(i)} \leq k\lambda,$$

where k is the least integer such that $\alpha^k > n/2$.

Proof. For any $i \in A$, we apply Lemma 6.7.1 to find a sequence $S = \{i_1, i_2, \ldots, i_{m+1}\}$ of points of A with $m \leq 2k$ and such that $i = i_1 = i_{m+1}$ and $(i_s, \tau(i_{s+1}))$ for all $1 \leq s \leq m$. We then define a new assignment σ (as suggested by Figure 6.3) by first taking

$$\sigma(i_s) = \tau(i_{s+1}) \quad \text{for all } 1 \leq s \leq m$$

and then handling the rest of the assignment by copying τ,

$$\sigma(j) = \tau(j) \quad \text{for all } j \notin S.$$

By the suboptimality of σ, we have

$$\sum_{j \in A} w_{j\tau(j)} \leq \sum_{j \in A} w_{j\sigma(j)},$$

and we can cancel the terms with $j \notin S$, so we have the bound

(6.25) $$\sum_{j \in S} w_{j\tau(j)} \leq \sum_{j \in S} w_{j\sigma(j)} \leq \lambda \,\text{card}\, S.$$

Since $i \in S$ by construction, $w_{i\tau(i)}$ is one of the summands in the left-hand sum in (6.25). Since card $S \leq k$, the proof of the lemma is therefore complete.

We now need to check that if we take all of the edges with c_{ij} less than a suitable multiple of $\log n/n$, then with high probability we have an expander graph. Such a result is given as Lemma 8 of Karp and Steele (1985), but the following version is more precise.

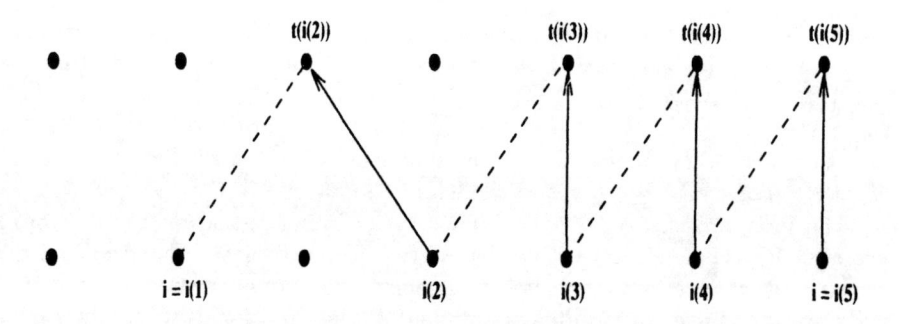

FIG. 6.3. *Dotted lines give sigma in terms of tau.*

PROPOSITION 6.7.2. *Let G' be the subgraph of G obtained by taking the edge set $E(G') = \{(i,j) : c_{ij} \leq 2u \log n/n\}$. There is a constant $K > 0$ not depending on n or u such that for all $u > K$, the graph G' is α-expanding for $\alpha = u \log n$ with probability at least $1 - n^{-u/K}$.*

Proof. For any $S \subset \{1, 2, \ldots, n\}$ with card $S = s$, we consider the set of events $A_j = \{j \notin \Gamma'(S)\}$, where $\Gamma'(S)$ denotes the set of neighbors of S in G'. We first note that

$$(6.26) \qquad P(A_j) = (1 - 2u \log n/n)^s \leq \exp(-2su \log n/n),$$

and if B_i is the complement of A_i, we have

$$\operatorname{card} \Gamma'(S) = \sum_{i=1}^{n} 1_{B_i}.$$

Moreover, the events B_i are all independent since they are functions of disjoint subsets of the c_{ij}'s. By the usual moment-generating-function argument, one can easily show that for any $0 \leq \delta < 1$ there is a $\phi(\delta) > 0$ such that for any collection of independent events B_i, we have

$$(6.27) \qquad P\left(\sum_i 1_{B_i} \leq \delta \sum_i P(B_i)\right) \leq \exp\left(-\phi(\delta) \sum_i P(B_i)\right).$$

We therefore need a lower bound on $\rho = \sum P(B_i) = n(1 - \exp(-2su \log n/n))$. The elementary bound $1 - e^{-x} \geq x(1 - e^{-1})$ for $0 \leq x \leq 1$ gives us

$$P(B_i) \geq (1 - e^{-1}) 2su \log n/n.$$

For s such that $\alpha s < n/2$, we have

$$P(\Gamma'(S) < (1 - e^{-1}) 2\alpha s) \leq \exp\left(-\rho \phi\left(\frac{s\alpha}{\rho}\right)\right).$$

By our estimate of ρ, we also have $s\alpha/\rho \leq 2^{-1}(1 - e^{-1})^{-1} \equiv \delta < 1$. Now we have the final calculation,

$$P(\exists S \text{ with card } S \leq n/2\alpha \text{ such that } \Gamma'(S) \leq \alpha \operatorname{card} S)$$

$$\leq \sum_s \binom{n}{s} \exp(-su \log n \phi(\delta)) \leq \exp(-u \log n/K),$$

where in the last step we applied the trivial bound n^s on the binomial coefficient and then summed the geometric series. This proves the first condition needed for G' to be an α expander graph; namely,

$$\alpha \operatorname{card} S \leq n/2 \to \operatorname{card} \Gamma'(S) \geq \alpha \operatorname{card} S.$$

One can prove the second condition by a similar calculation.

Now we just collect the consequences of our analysis. The formal statement looks more involved than one might hope, but the bounds are the natural ones given the tools that have been used. These inequalities are not likely to be sharp, but if one keeps in mind that no useful bound on the concentration of A_n was available before Talagrand (1995), the progress is substantial. In the simplest terms, the theorem tells us that (up to a logarithmic factor) the standard deviation of A_n is of order $O(n^{-\frac{1}{2}})$.

THEOREM 6.7.1 (Talagrand (1995)). *If m_n is the median of A_n, then for $n \geq 3$ we have for $t \leq \sqrt{\log n}$ that*

$$P\left(|A_n - m_n| \geq \frac{Kt(\log n)^2}{\sqrt{n}\log\log n}\right) \leq 2\exp(-t^2),$$

and for $t \geq \sqrt{\log n}$ we have that

$$P\left(|A_n - m_n| \geq \frac{Kt^3(\log n)}{\sqrt{n}(\log t)^2}\right) \leq 2\exp(-t^2).$$

Proof. The first step just requires that we recall that with high probability A_n will require no edges that are much more expensive than $(\log n)^\beta / n$. To set this out more honestly, we take $u \leq n/(2\log n)$ and $\alpha = u\log n$ and let m be the smallest integer such that $\alpha^m \geq n/2$. If we take $v = 4mu\log n/n$, then with $c'_{ij} = \min(c_{ij}, v)$, the cost A_n^u of the minimal assigment under the costs c'_{ij} satisfies

$$P(A_n = A_n^u) \geq 1 - n^{-u/K}$$

because of Proposition 6.7.1 and the fact that the graph G' with edges $\{(ij) : c_{ij} \leq v\}$ is α-expanding with probability at least $1 - n^{-u/K}$.

The next step uses our large-deviation bound for suprema of linear functionals. After scaling, we find that

$$P(|A_n - m_n| \geq x) \leq 2\exp(-x^2/(4nv^2));$$

so, all told, we have the bound

$$P(|A_n - m_n| \geq x) \leq n^{-u/K} + 2\exp(-x^2/(4nv^2)).$$

The only remaining task is to make appropriate choices of our free parameters. If we first set $x = 3tvn^{\frac{1}{2}}$, then for $t \leq \sqrt{\log n}$ we cannot do better than to take $u = K$, while for $t \geq \sqrt{\log n}$ calculus suggests the choice $u = Kt^2/\log n$. These selections provide the proof of the theorem.

6.8. Further applications of Talagrand's isoperimetric inequalities.

Talagrand's convex-distance inequality is too new for one to be able to say what constitutes a typical application, but the applications that have been given here certainly provide a pattern that can be repeated many times. Certainly, the method used to prove the Gaussian tail bound for the TSP and Steiner problems can be used with small variations on many other problems—though naturally, in any specific problem, there may be difficulties that must be overcome. For instance, to obtain a Gaussian tail bound for the length of the minimal spanning tree, one must come to grips with the lack of monotonicity of the MST functional. Still, the convex-distance inequality almost always provides a powerful place from which to begin the analysis, and for the MST it also prevails.

All of the examples that have been studied in this chapter have called on just one of Talagrand's isoperimetric inequalities, the one for the convex-distance inequality. This biased selection helped the exposition to be more cohesive, but this convenience comes at a price. In the discussion of the proof of Theorem 6.1.1, it was noted that the technique used there to carry out an induction over the dimensions is remarkably general, and Theorem 6.1.1 is just one particular consequence of the technique. To obtain an honest view of the variety of isoperimetic inequalities that can be obtained by the induction technique, one naturally needs to study the seminal paper Talagrand (1995).

6.9. Final considerations on related work.

Many important topics that have a close relationship to this monograph have had to be omitted. Perhaps the most obvious omissions concern the theory of random graphs and their associated algorithms. Many results from that theory would fit nicely into the present development, but as a matter of design almost all such results have been omitted. Fortunately, the volumes of Bollobás (1985) and Alon, Spencer, and Erdös (1992) can provide the reader with a delightful view of this important field.

A second important ommission is the theory of Markov-chain Monte Carlo. This area—which includes the methods of simulated annealing, probabilitic enumeration, and the Gibbs sampler—is undergoing explosive development. The methods have great practical impact, and they have stimulated many subtle methodological developments. Still, the state of flux in this field is probably too great to permit any one volume to do justice to the activity. Thus, by design, all results from this development have been omitted from this monograph. The reader should find that the volume of Sinclair (1993) provides an excellent starting point for the voyage into the part of Markov-chain Monte Carlo that is closest in spirit to the problems that have been studied here.

Perhaps the most regrettable omission from the present volume is the discussion of the two-sample matching problem. To give a glimpse into that theory, we first consider two independent samples $\{X_1, X_2, \ldots, X_n\}$ and $\{Y_1, Y_2, \ldots, Y_n\}$ that are both uniformly distributed on the unit square $[0, 1]^2$. The probability theory of the two-sample matching problem begins with the consideration of the

random variable
$$M_n = \min_\sigma \sum_{i=1}^n |X_i - Y_{\sigma(i)}|.$$

Here, as ususal, the minimum is over all permuations, and $|x - y|$ denotes the Euclidean distance from x to y. Many of the methods that have been developed in this volume might seem to apply off the shelf to provide an asymptotic understanding of M_n, but M_n is filled with surprises. The first of these was the discovery of Ajtai, Komlós, and Tusnády (1984) that there are constants c_0 and c_1 such that

$$c_0 \sqrt{n \log n} \leq EM_n \leq c_1 \sqrt{n \log n}.$$

This result tells us that EM_n does not exhibit the typical $\sqrt{n}$ behavior that we have come to expect from results that are related to the Beardwood–Halton–Hammersley theorem. Moreover, as a measure of the subtle nature of M_n, we should note that it is still an open problem to show that $EM_n \sim c_0 \sqrt{n \log n}$ as $n \to \infty$.

The appearance of the extra $\log n$ in the analysis of M_n is an unexpected twist, but by no means is it the last twist one finds in the theory of two-sample matching. For example, suppose that we consider the matching which does the best job in the sense of minimizing the longest edge in the matching. In other words, we consider the random variable defined by

$$M_N^* = \min_\sigma \max_{1 \leq i \leq n} |X_i - Y_{\sigma(i)}|.$$

Quite remarkably, Leighton and Shor (1989) proved that we have

$$c_0 n^{-1/2} (\log n)^{3/4} \leq EM_n^* \leq c_1 n^{-1/2} (\log n)^{3/4}.$$

Here we have a mysterious factor of $(\log n)^{\frac{3}{4}}$ for which one is at first—and perhaps for a good while—hard pressed to find an intuitive explaination.

In addition to the obvious intellectual appeal of the two-sample matching problem, there are important applications of the theory. One can find a full discussion of these applications and many related results in the volume of Coffman and Leuker (1991).

The story of the matching variables turns out to be exceptionally rich, and a key aspect of the theory of these variables is that they are closely connected to the theory of empirical processes. In fact, by leaning strongly on earlier developments in the theory of empirical processes, Talagrand (1994) has provided a view of M_n and M_n^* that unifies their analysis and provides a deep understanding of their behavior in terms of the notion of majorizing measures.

Bibliography

Adler, R. (1986), Commentary in *The Collected Works of S. Kakutani, Vol. II*, R. R. Kalman, ed., Birkhauser, Boston, 1986, p. 444.

Aho, A. V. (1990), "Algorithms for finding patterns in strings," in *Handbook of Theoretical Computer Science*, J. van Leeuwen, ed., North–Holland, Amsterdam, pp. 256–300.

Ajtai, M., Komlós, J., and Tusnády, G. (1984), "On optimal matchings," *Combinatorica*, 4, pp. 259–264.

Aldous, D. (1989), *Probability Approximations via the Poisson Clumping Heuristic*, Springer-Verlag, New York.

Aldous, D. (1990), "A random tree model associated with random graphs," *Random Structures Algorithms*, 1, pp. 383–402.

Aldous, D. (1992), "Asymptotics in the random assignment problem," *Probab. Theory Appl.*, 93, pp. 507–534.

Aldous, D. and Steele, J. M. (1992), "Asymptotics for Euclidean minimal spanning trees on random points," *Probab. Theory Related Fields*, 92, pp. 247–258.

Alexander, K. S. (1993), "A note on rates of convergence in first passage percolation," *Ann. Appl. Probab.*, 3, pp. 81–90.

Alexander, K. S. (1994a), "Rates of convergence for distance-minimizing subadditive Euclidean functionals," *Ann. Appl. Probab.*, 4, pp. 902–922.

Alexander, K. S. (1994b), "The rate of convergence of the mean of the longest common subsequence," *Ann. Appl. Probab.*, 4, pp. 1074–1083.

Alexander, K. S. (1996), "The RSW Theorem for continuum percolation and the CLT for the Euclidean spanning tree," *Ann. Appl. Probab.*, 6, pp. 466–494.

Alexander, K. S. (1995a), "Simultaneous uniqueness of infinite clusters in stationary random labeled graphs," to appear in *Comm. Math. Phys.*

Alexander, K. S. (1995b), "Percolation and minimal spanning forests in infinite graphs," *Ann. Probab.*, 23, pp. 902–922.

Alexander, K. S. (1995c), "Approximation of subadditive functions and convergence rates in limiting-shape results," to appear in *Ann. Probab.*

Alon, N., Spencer, J. H., and Erdös, P. (1992), *The Probabilistic Method*, Wiley-Interscience Series in Discrete Mathematics and Optimization, John Wiley, New York.

Alpern, S. (1977), "Dyadic decompostitions of the cube," in *Proc. Sixth Southeast Conference in Combinatorics, Graph Theory and Computing*, Utilitas Mathematica Publishing, Winnepeg, pp. 47–56.

Apostolico, A. and Guerra, C. (1987), "The longest common subsequence problem revisited," *Algorithmica*, 2, pp. 315–332.

Avis, D. and Devroye, L. (1985), "An analysis of a decomposition heuristic for the assignment problem," *Oper. Res. Lett.*, 3, pp. 279–283.

Avis, D. and Lai, C. W. (1988), "The probabilistic analysis of a heuristic for the assignment problem," *SIAM J. Comput.*, 17, pp. 732–741.

Avram, F. (1991), "Asymptotic behavior of the Euclidean traveling salesman with large tails," Technical report, Northeastern University, Boston.

Avram, F. and Bertsimas, D. (1992), "The minimal spanning tree constant in geometric probability and under the independent model: A unified approach," *Ann. Appl. Probab.*," 2, pp. 113–130.

Avram, F. and Bertsimas, D. (1993), "On central limit theorems in geometrical probability," *Ann. Appl. Probab.*, 3, pp. 1033–1046.

Azuma, K. (1967), "Weighted sums of certain dependent random variables," *Tôhoku Math. J.*, 19, pp. 357–367.

Bailey, T. (1969), "Spacefilling curves: Their generation and their application to bandwidth reduction," *IEEE Trans. Inform. Theory*, II-15, pp. 658–664.

Baldi, P. and Rinott, Y. (1989), "On normal approximations of distributions in terms of dependency graphs," *Ann. Probab.*, 17, pp. 1646–1650.

Beardwood, J., Halton, J. H., and Hammersley, J. M. (1959), "The shortest path through many points," *Proc. Cambridge Philos. Soc.*, 55, pp. 299–327.

Bellman, R. E. (1962), "Dynamic programming treatment of the traveling salesman problem," *J. Assoc. Comput. Mach.*, 9, pp. 61–63.

Bern, M. and Epstein, D. (1993), "Worst case bounds for subadditive geometric graphs," in *Proc. Ninth Annual Symposium on Computational Geometry*, Association for Computing Machinery, New York, pp. 183–188.

Bertsekas, D. P. (1991), *Linear Network Optimization: Algorithms and Codes*, MIT Press, Cambridge, MA.

Bickel, P. J. and Breiman, L. (1983), "Sums of functions of nearest neighbor-distances, moment bounds, limit theorems, and goodness of fit test," *Ann. Statist.*, 11, pp. 186–214.

Bingham, N. H. (1981), "Tauberian theorems and the central limit theorem," *Ann. Probab.*, 9, pp. 221–231.

Bollobás, B. (1985), *Random Graphs*, Academic Press, New York.

Bollobás, B. (1987), "Martingales, isoperimetric inequalities, and random graphs," *Colloq. Math. Soc. János Bolyai*, 52, pp. 113–139.

Bollobás, B. (1990), "Sharp concentration of measure phenomena in random graphs," in *Random Graphs*, M. Karonski, J. Jawarski, and A. Rucinski, eds., Ann. Discrete Math. Ser., John Wiley, New York, pp. 1–15.

Bollobás, B. and Leader, I. (1991a), "Compression and isoperimetric inequalities," *J. Combin. Theory* A, 56, pp. 47–62.

Bollobás, B. and Leader, I. (1991b), "Isoperimetric inequalities and fractional set system," *J. Combin. Theory* A, 56, pp. 63–74.

Burkholder, D. L. (1966), "Martingale transforms," *Ann. Math. Statist.*, 37, pp. 1494–1504.

Burkholder, D. L. (1973), "Distribution funciton inequalities for martingales," *Ann. Probab.*, 1, pp. 19–42.

Chvátal, V. and Sankoff, D. (1975), "Longest common subsequences of two random sequences," *J. Appl. Probab.*, 12, pp. 306–315.

Chvátal, V. (1980), *Linar Programming*, W. H. Freeman, San Francisco.

Coffman, E. G., Jr. and Leuker, G. S. (1991), *Probabilistic Analysis of Packing and Partitioning Algorithms*, John Wiley, New York.

Dančík, V. and Paterson, M. (1994), "Upper bound for expected length of a longest common subsequence of two random sequences," in *Proc. STACS '94*, Lecture Notes in Computer Science 775, Springer-Verlag, New York, pp. 669–678.

Davis, B., Avis, D., and Steele, J. M. (1988), "Probabilistic analysis of a greedy heuristic for Euclidean matching," *Probab. Engrg. Inform. Sci.*, 2, pp. 143–156.

DeBruijn, N. G. and Erdös, P. (1952a), "Some linear and some quadratic recursion formulas I," *Indag. Math.*, 13, pp. 374–382.

DeBruijn, N. G. and Erdös, P. (1952b), "Some linear and some quadratic recursion formulas II," *Indag. Math.*, 14, pp. 152–163.

Deken, J. P. (1979), "Some limit results for longest common subsequences," *Discrete Math.*, 26, pp. 17–31.

Donath, W. E. (1969), "Algorithm and average-value bounds for assignment problems," *IBM J. Res. Develop.*, 13, pp. 380–386.

Dyer, M. E., Frieze, A. M., and McDiarmid, C. J. H. (1986), "On linear programs with random costs," *Math. Programming*, 35, pp. 3–16.

Efron, B. and Stein, C. (1981), "The jackknife estimate of variance," *Ann. Statist.*, 9, pp. 586–596.

Erdös, P. and Szekeres, G. (1935), "A combinatorial problem in geometry," *Composito Math.*, 2, pp. 463–470.

Erdös, P. and Lovász, L. (1975), "Problems and results on 3-chromatic hypergraphs and some related questions," in *Infinite and Finite Sets*, A. Hajnal et al., eds., North–Holland, Amsterdam, pp. 609–628.

Fekete, M. (1923), "Über die Verteilung der Wurzeln bei gewissen algebraischen Gleichungen mit ganzzähligen Koeffizienten," *Math. Z.*, 17, pp. 228–249.

Few, L. (1962), "Average distances between points in a square," *Mathematika*, 9, pp. 111–114.

Fredman, M. L. (1975), "On computing the length of longest increasing subsequences," *Discrete Math.*, 11, pp. 29–35.

Frenk, J. B. G., van Houweninge, M. V., and Rinnooy Kan, A. H. G. (1987), "Order statistics and the linear assignment problem," *Computing*, 39, pp. 165–174.

Frieze, A. M. (1991), "On the length of the longest monotone increasing subsequence in a random permutation," *Ann. Appl. Probab.*, 1, pp. 301–305.

Garsia, A. M. (1976), "Combinatorial inequalities and smoothness of functions," *Bull. Amer. Math. Soc.*, 82, pp. 157–170.

Gao, J. and Steele, J. M. (1994a), "Sums of squares of edge lengths and spacefilling curve heuristics for the traveling salesman problem," *SIAM J. Discrete Math.*, 7, pp. 314–324.

Gao, J. and Steele, J. M. (1994b), "General spacefilling curve heuristics and limit theory for the traveling salesman problem," *J. Complexity*, 10, pp. 230–245.

Goddyn, L. (1990), "Quantizers and the worst-case Euclidean traveling salesman problem," *J. Combin. Theory* B, 50, pp. 65–81.

Goemans, M. X. and Kodialam, M. S. (1993), "A lower bound on the expected cost of an optimal assignment," *Math. Oper. Res.*, 18, pp. 267–274.

Groeneboom, P. (1988), "Limit theorems for convex hulls," *Probab. Theory Related Fields*, 79, pp. 329–369.

Halton, J. H. and Terada, R. (1982), "A fast algorithm for the Euclidean traveling salesman problem, optimal with probability one," *SIAM J. Comput.*, 11, pp. 28–46.

Hammersley, J. M. (1962), "Generalization of the fundamental theorem of subadditive functions," *Proc. Cambridge Philos. Soc.*, 58, pp. 235–238.

Hammersley, J. M. (1972), "A few seedlings of research," in *Proc. Sixth Berkeley Sympos. Math. Statist. Probab.*, University of California Press, Berkeley, CA, pp. 345–394.

Hammersley, J. M. (1974), "Postulates for subadditive processes," *Ann. Probab.*, 2, pp. 652–680.

Heuter, I. (1994), "On the convex hull of a random sample," *Adv. Appl. Probab.*, 26, pp. 855–875.

Hilbert, D. (1891), "Über die stetige Abbildung einer Linie auf ein Flächenstück," *Math. Ann.*, 38, pp. 459–460.

Hille, E. and Phillips, R. S. (1957), *Functional Analysis and Semi-Groups*, American Mathematical Society, Providence, RI.

Hirschberg, D. S. (1977), "Algorithms for the longest common subsequence problem," *J. Assoc. Comput. Mach.*, 24, pp. 664–675.

Hochbaum, D. and Steele, J. M. (1982), "Steinhaus' geometric location problem for random samples in the plane," *Adv. Appl. Probab.*, 14, pp. 55–67.

Hoeffding, W. (1963), "Probability inequalities for sums of bounded random variables," *J. Amer. Statist. Assoc.*, 58, pp. 13–30.

Hunt, J. W. and Syzmansky, T. G. (1977), "A fast algorithm for computing the longest common subsequences," *Comm. Assoc. Comput. Mach.*, 20, pp. 350–353.

Hsing, T. (1994), "On the asymptotic distribution of the area outside a random convex hull in the disk," *Ann. Appl. Probab.*, 4, pp. 478–493.

Imai, H. (1986), "Worst case analysis for planar matching and tour heuristics with bucketing techniques and spacefilling curves," *J. Oper. Res. Soc. Japan*, 29, pp. 43–67.

Jaillet, P. (1992), "Rates of convergence for quasi-additive smooth Euclidean functionals and applications to combinatorial optimization problems," *Math. Oper. Res.*, 17, pp. 965–980.

Jaillet, P. (1993), "Rate of convergence for the Euclidean minimum spanning tree law," *Oper. Res. Lett.*, 14, pp. 73–78.

Kahane, J.-P. (1976), "Mesure et dimensions", in *Turbulence and Navier Stokes Equations*, A. Dold and B. Eckmann, eds., Lecture Notes in Mathematics 565, Springer-Verlag, New York.

Karlin, S. and Rinott, Y. (1982), "Application of ANOVA type decompositions for comparison of variance statistics including jackknife estimates," *Ann. Statist.*, 10, pp. 485–501.

Karloff, H. J. (1989), "How long can a traveling salesman tour be?", *SIAM J. Discrete Math.*, 2, pp. 91–99.

Karp, R. M. (1976), "The probabilistic analysis of some combinatorial search algorithms," in *Algorithms and Complexity: New Directions and Recent Results*, J. F. Traub, ed., Academic Press, New York, pp. 1–19.

Karp, R. M. (1977), "Probabilistic analysis of partitioning algorithms for the traveling-salesman problem in the plane," *Math. Oper. Res.*, 2, pp. 209–224.

Karp, R. M. (1979), "A patching algorithm for the nonsymmetric traveling-salesman problem," *SIAM J. Comput.*, 8, pp. 561–573.

Karp, R. M. (1983), Lecture at the NIHE Summer School on Combinatorial Optimization, Dublin, unpublished.

Karp, R. M. (1987), "An upper bound on the expected cost of an optimal assignment," in *Discrete Algorithms and Complexity: Proceedings of the Japan-US Joint Seminar*, D. Johnson et al., eds., Academic Press, New York, pp. 1–4.

Karp, R. M., Rinnooy Kan, A. H. G., and Vohra, R. (1994), "Average case analysis of a heuristic for the assignment problem," *Math. Oper. Res.*, 19, pp. 513–522.

Karp, R. M. and Steele, J. M. (1985), "Probabilistic analysis of heuristics," in *The Traveling Salesman Problem: A Guided Tour of Combinatorial Optimization*, E. L. Lawler, et al., eds., John Wiley, New York, pp. 181–206.

Karp, R. M., Lenstra, J. K., McDiarmid, C. J. H., and Rinnooy Kan, A. H. G. (1984), "Probabilistic analysis of combinatorial algorithms: An annotated bibliography," in *Combinatorial Optimization: Annotated Bibliographies*, M. O. Óh Éigeartaigh, J. K. Lenstra, and A. G. H. Rinnooy Kan, eds., Wiley-Interscience, New York, pp. 52–88.

Kesten, H. (1973), Contribution to J. F. C. Kingman, "Subadditive ergodic theory," *Ann. Probab.*, 1, pp. 883–909.

Kesten, H. and Lee, S. (1996), "The central limit theorem for weighted minimal spanning trees on points," *Ann. Appl. Probab.*, 6, pp. 495–527.

Kingman, J. F. C. (1968), "The ergodic theory of subadditive stochastic processes," *J. Roy. Statist. Soc. Ser.* B, 30, pp. 499–510.

Kingman, J. F. C. (1973), "Subadditive ergodic theory," *Ann. Probab.*, 1, pp. 883–909.

Kurtzberg, J. M. (1962), "On approximation methods for the assignment problem," *J. Assoc. Comput. Mach.*, 9, pp. 419–439.

Lazarus, A. (1979), "The assignment problem with uniform (0, 1) cost matrix," Unpublished B. A. thesis, Department of Mathematics, Princeton University, Princeton, NJ.

Leader, I. (1991), "Discrete isoperimetric inequalities," *Proc. Sympos. Appl. Math.*, 44, pp. 57–80.

Leighton, T. and Shor, P. (1989), "Tight bounds for minimax grid matching with applications to the average case analysis of algorithms," *Combinatorica*, 9, pp. 161–187.

Lindvall, T. (1992), *The Coupling Method*, John Wiley, New York.

Lin, S. and Kernigan, B. W. (1973), "An effective heuristic algorithm for the traveling salesman problems," *J. Oper. Res.*, 21, pp. 498–516.

Logan, B. F. and Shepp, L. A. (1977), "A variational problem for Young tableaux," *Adv. Math.*, 26, pp. 206–222.

Meyers, E. W. (1986), "An $O(ND)$ difference algorithm and its variations," *Algorithmica*, 1, pp. 251–266.

Maurey, B. (1979), "Construction de suites symétriques." *Comptes Rendu Acad. Sci. Paris*, 288, pp. 679–681.

McDiarmid, C. J. H. (1986), "On the greedy algorithm with random costs," *Math. Programming*, 36, pp. 245–255.

McDiarmid, C. J. H. (1989), "On the method of bounded differences," in *Surveys in Combinatorics: London Mathematical Society Leture Notes*, 141, Cambridge University Press, Cambridge, UK, pp. 148–188.

Mézard, M. and Parisi, G. (1987), "On the solution of the random link matching problem," *J. Physique*, 48, pp. 1451–1459.

Mézard, M. and Parisi, G. (1988), "The Euclidean matching problem," *J. Physique*, 49, pp. 2019–2025.

Milman, V. D. and Schechtman, G. (1986), *Asymptotic Theory of Finite Dimensional Normed Spaces*, Lecture Notes in Mathematics 1200, Springer-Verlag, New York.

Milne, S. C. (1980), "Peano curves and smoothness of functions," *Adv. Math.*, 35, pp. 129–157.

Myers, E. W. (1986), "An $O(ND)$ difference algorithm and its variations," *Algorithmica*, 1, pp. 251–266.

Papadimitriou, C. H. (1978a), "The Euclidean traveling salesman problem is NP-complete," *Theoret. Comput. Sci.*, 4, pp. 237–244.

Papadimitriou, C. H. (1978b), "The probabilistic analysis of matching heuristics," in *Proc. 15th Annual Conference Comm. Control Computing*, University of Illinois at Urbana–Champaign, Urbana, IL, pp. 363–378.

Peano, G. (1890), "Sur une courrbe qui remplit toute une aire plane," *Math. Ann.*, 36, pp. 157–160.

Penrose, M. D. (1995a), "Continuum percolation and Euclidean minimal spanning trees in high dimensions," *Ann. Appl. Probab.*, 6, pp. 528–544.

Penrose, M. D. (1995b), "The random minimal spanning tree in high dimensions," Technical report, Department of Mathematics, Durham University, Durham, UK.

Platzman, L. K. and Bartholdi, J. J. (1989), Spacefilling curves and the planar traveling salesman problem, *J. Assoc. Comput. Mach.*, 36, pp. 719–737.

Ramey, D. B. (1982), "A non-parametric test of bimodality with applications to cluster analysis," Doctoral dissertation, Department of Statistics, Yale University, New Haven, CT.

Redmond, C. and Yukich, J. E. (1994), "Limit theorems and rates of convergence for Euclidean functionals," *Ann. Appl. Probab.*, 4, pp. 1057–1073.

Rényi, A. and Sulanke, R. (1963), "Über die konvexe Hülle von n zufällig gewählten Punkten I," *Z. Wahrscheinlichkeitstheorie und verwandte Gebiete*, 2, pp. 75–84.

Rényi, A. and Sulanke, R. (1964), "Über die konvexe Hülle von n zufällig gewählten Punkten II," *Z. Wahrscheinlichkeitstheorie und verwandte Gebiete*, 3, pp. 138–147.

Rhee, W. T. (1992), "On the travelling sales person problem in many dimensions," *Random Structures Algorithms*, 3, pp. 227–233.

Rhee, W. T. (1993a), "A matching problem and subadditive Euclidean functionals," *Ann. Appl. Probab.*, 3, pp. 794–801.

Rhee, W. T. (1993b), "On the stochastic Euclidean traveling salesperson problem for distributions with unbounded support," *Math. Oper. Res.*, 18, pp. 292–299.

Rhee, W. T. (1994), "Boundary effects in the traveling salesperson problem," *Oper. Res. Lett.*, 16, pp. 19–25.

Rhee, W. T. (1995), "On rates of convergence for common subsequences and first passage time," *Ann. Appl. Probab.*, 5, pp. 44–48.

Rhee, W. T. and Talagrand, M. (1987a), "Martingale inequalities NP-complete problems," *Math. Oper. Res.*, 12, pp. 177–181.

Rhee, W. T. and Talagrand, M. (1989a), "Martingale inequalities, interpolation and NP-complete problems," *Math. Oper. Res.*, 14, pp. 91–96.

Rhee, W. T. and Talagrand, M. (1989b), "A sharp deviation for the stochastic traveling salesman problem," *Ann. Probab.*, 17, pp. 1–8.

Richardson, D. (1973), "Random growth in a tesselation," *Proc. Cambridge Philos. Soc.*, 74, pp. 515–528.

Rinsma-Melchert, I. (1993), "The expected number of matches in optimal global sequence alignments," *New Zealand J. Botany*, 31, pp. 219–230.

Samuels, S. and Steele, J. M. (1981), "Optimal sequential selection of a monotone sequence from a random sample," *Ann. Probab.*, 9, pp. 937–947.

Sankoff, D. and Kruskal, J. B., eds. (1983), *Time Warps, String Edits, and Macromolecules: The Theory and Practice of Sequence Comparison*, Addison–Wesley, Reading, MA.

Shamir, E. and Spencer, J. (1987), "Sharp concentration of the chromatic number on random graphs $G_{n,p}$," *Combinatorica*, 7, pp. 121–129.

Shields, P. (1987), "The ergodic and entropy theorems revisited," *IEEE Trans. Inform. Theory*, IT-33, pp. 263–265.

Sinclair, A. (1993), *Algorithms for Random Generation and Counting: A Markov Chain Approach*, Birkhäuser, Boston.

Snyder, T. L. and Steele J. M. (1995a), "A priori bounds on the Euclidean traveling salesman," *SIAM J. Comput.*, 24, pp. 665–671.

Snyder, T. L. and Steele J. M. (1995b), "Equidistribution in all dimensions of worst-case point sets for the traveling salesman problem," *SIAM J. Discrete Math.*, 8, pp. 678–683.

Stadje, W. (1994), "Two asymptotic inequalities for the stochastic traveling salesman problem," Technical report, University of Osnabrück, Osnabrück, Germany.

Steele, J. M. (1981a), "Subadditive Euclidean functionals and non-linear growth in geometric probability," *Ann. Probab.*, 9, pp. 365–376.

Steele, J. M. (1981b), "Complete convergence of short paths and Karp's algorithm for the TSP," *Math. Oper. Res.*, 6, pp. 374–378.

Steele, J. M. (1982a), "Optimal triangulation of random samples in the plane," *Ann. Probab.*, 10, pp. 548–553.

Steele, J. M. (1982b), "Long common subsequences and the proximity of two random strings," *SIAM J. Appl. Math.*, 42, pp. 731–737.

Steele, J. M. (1986), "Probabilistic algorithm for the directed traveling salesman problem," *Math. Oper. Res.*, 11, pp. 343–350.

Steele, J. M. (1988), "Growth rates of Euclidean minimal spanning trees with power weighted edges," *Ann. Probab.*, 16, pp. 1767–1787.

Steele, J. M. (1989a), "Seedlings in the theory of shortest paths," in *Disorder in Physical Systems: A Volume in Honor of J. M. Hammersley*, G. Grimmett and D. Welsh, eds., Cambridge University Press, Cambridge, UK.

Steele, J. M. (1989b), "Kingman's subadditive ergodic theorem," *Ann. Inst. H. Poincaré Probab. Statist.*, 25, pp. 93–98.

Steele, J. M. (1995), "Variations on the long increasing subsequence theme of Erdös and Szekeres," in *Discrete Probability and Algorithms*, D. Aldous,

P. Diaconis, and J. M. Steele, eds., Volumes in Mathematics and Its Applications 72, Springer-Verlag, New York, pp. 111–131.

Steele, J. M., Shepp, L. A., and Eddy, W. F. (1987), "On the number of leaves of a Euclidean minimal spanning tree," *J. Appl. Probab.*, 24, pp. 809–826.

Stein, C. (1986), *Approximate Calculation of Expectations*, IMS Monograph Series, Institute of Mathematical Statistics, Hayward, CA.

Talagrand, M. (1988), "An isoperimetric theorem on the cube and the Khintchine–Kahane inequalities," *Proc. Amer. Math. Soc.*, 104, pp. 905–909.

Talagrand, M. (1992), "Matching random samples in many dimensions," *Ann. Appl. Probab.*, 2, pp. 846–856.

Talagrand, M. (1994), "Matching theorems and empirical discrepancy computations using majorizing measures," *J. Amer. Math. Soc.*, 7, pp. 455–537.

Talagrand, M. (1995), "Concentration of measure and isoperimetric inequalities in product spaces," *Publ. Math. IHES*, 81, pp. 73–205.

Yao, A. (1991), "Probabilistic behavior of shortest paths over unbounded regions," manuscript.

Yukich, J. E. (1995), "Asymptotics for the Euclidean TSP with power weighted edges," *Probab. Theory Related Fields*, 102, pp. 203–220.

Yukich, J. E. (1996a), "Worst case growth rates for some classical optimization problems," *Combinatorica*, to appear.

Yukich, J. E. (1996b), "Ergodic theorems for some classical problems in combinatorial optimization," *Ann. Appl. Probab.*, to appear.

Veršik, A. M. and Kerov, C. V. (1977), "Asymptotics of the Plancherel measure of the symmetric group and a limiting form for Young tableaux," *Dokl. Akad. Nauk USSR*, 233, pp. 1024–1027.

Walkup, D. W. (1979), "On the expected value of a random assignment problem," *SIAM J. Comput.*, 8, pp. 440–442.

Walkup, D.W. (1981), "Matchings in random regular bipartite digraphs," *Discrete Math.*, 31, pp. 59–64.

Waterman, M. S. (1984), "General methods of sequence comparison," *Bull. Math. Biol.*, 46, pp. 473–500.

Waterman, M. S. (1994), "Estimating statistical significance of sequence alignment," *Philos. Trans. Roy. Soc. London Ser.* B, 344, pp. 383–390.

Weide, B. W. (1978), "Statistical methods in algorithm design and analysis," Unpublished Ph.D. thesis, Department of Computer Science, Carnegie–Mellon University, Pittsburgh, PA.

Index

Adler, R., 49
Aho, A.V., 18
Ajtai, M., 141
Aldous, D., 25, 94, 96, 100, 117
Alexander, K. S., 13, 24, 67, 100, 108, 109
Alon, N., 24, 112
Alpern, S., 50
Alternating chain lemma, 136
Apostolico, A., 18
Assignment problem, 77, 135
    greedy algorithm, 77
    linear programming connection, 81
    matching theory method, 78
    Talagrand's tail bound, 139
Avis, D., 93
Avram, F., 49, 108, 110, 114, 118
Azuma's inequality, 5
Azuma, K., 4

Bailey, T., 49
Baldi, P., 112
Bartholdi, J. J., 43
Beardwood–Halton–Hammersley (BHH) theorem
    general case, 34
    uniform case, 33
Beardwood, J., 33
Bellman, R., 40
Bern, M., 117
Bersekas, D. P., 94

Bertsimas, D., 108, 110, 114, 118
Bickel, P., 110
Bingham, N. H., 25
Bland, R., 95
Bollobás, B., 24, 132
Breiman, L., 110
Burkholder's inequality, 45
Burkholder, D. L., 45

Central limit theory (CLT)
    conditioning methods, 110
    minimal spanning tree, 108
Chvátal, 1, 4, 94
Coffman, E. G., Jr., 141
Concentration inequality
    for the TSP, 29
    smooth functionals, 62
    space-filling curve heuristic, 44
Convex distance, 120

Dančík, V., 3
DeBruijn, N. G., 21
Deken, J. P., 3
Delaunay triangulation, 95
Devroye, L., 93
Doob, J., 5
Dyer–Frieze–McDiarmid inequality, 84
Dyer, M. E., 82
Dynamic programming, 16

Eddy, W. F., 99
Efron, B., 124

Eggert, M., 3
Epstein, D., 117
Erdös, P., 6, 21, 24, 112
Erdös–Szekeres theorem, 6, 17
Euclidean functionals, 53
Expander graphs, 137

Fekete, M., 2
Few, L., 50
Flipping lemma, 10
Flipping method, 10, 127
Fredman, M. L., 17
Frenk, J. B. G., 93
Frieze, A. M., 11, 13, 82

Gao, J., 43
Garsia, A., 49
Gaussian tail bound
   for the TSP, 124
Geometric subadditivity, 54, 60
Goddyn, L., 50
Goemans, M. X., 94
Groneboom, P., 116
Guerra, C., 18

Halton, J. H., 33, 41
Hammersley, J. M., 6, 26, 33
Hausdorff distance, 132
Hereditary sets, 131
Heurter, I., 116
Hilbert, D., 44
Hille, E., 25
Hirshberg, D. S., 18
Hochbaum, D., 25
Hoeffding, V., 4
Hsing, T., 116
Hunt, J. W., 18

Imai, H., 49
Increasing-subsequence problem,
   6, 75, 125
   concentration inequalities,
      10, 125
   sequential selection, 24
Isoperimetric inequality, 119
   Hamming distance, 61

Jaillet, P., 65, 72

Kahane, J.-P., 49
Kakutani, S., 49
Karloff, H. J., 50
Karp's partitioning algorithm, 40
Karp, R. M., 35, 40, 41, 50, 91,
   93, 137
Kerov, C. V., 10
Kesten, H., 76, 109, 117
Kingman, J. F. C., 18
Kodialam, M. S., 94
Komlós, J., 141
Kruskal, J. B., 24
Kurtzburg, J. M., 78

Lai, C. W., 93
Lazarus, A., 94
Leader, I., 132
Lee, S., 109, 117
Leighton, T., 141
Lens geometry for the MST, 107
Leuker, G. S., 141
Lindvall, T., 36
Logan, B., 10
Long-common-subsequence
      problem, 1
   algorithms, 18
   tail bound, 6
Lovász, L., 112

Markov-chain Monte Carlo, 140
Maurey, B., 24
McDiarmid, C. J. H., 24, 82, 93
McKay, D. B., 3
Meyers, E. W., 18
Mézard, M., 76, 94
Milman, V. D., 24
Milne, S. C., 49
Minimal-matching problem, 63
   two samples, 141
Minimal spanning tree (MST),
   97
   central limit theory, 109
   degree theorem, 99
   power-weighted edges, 99

Objective method, 96

INDEX

Papadimitriou, C., 40, 76
Parisi, G., 76, 94
Paterson, M., 3
Peano, G., 44
Phillips, R. S., 25
Platzman, L. K., 43
Poissonization device, 25
Pólya, G., 25

Ramey, D. B., 108, 109
Random graphs, 140
Rates of convergence
    long-common-subsequence
      problem, 13
    MST, 72
    rooted-dual method, 70
    TSP means, 67
Redmond, C., 68
Rényi, A., 116
Rhee, W. T., 14, 24, 33, 38, 59, 65, 72, 75, 124
Richardson, D., 76
Rinnooy Kan, A. H. G., 93
Rinnot, Y., 112
Rinsma-Melchert, I., 3
Rooted dual, 68
    for the minimal matching, 70
    for the MST, 70
    for the TSP, 68

Sankoff, D., 1, 4, 24
Schechtman, G., 24
Self-similarity, 44
Shamir, E., 24
Shepp, L. A., 10, 99
Shor, P., 141
Simple subadditivity, 60
Simplex method, 82
Snyder, T. L., 50, 117
Space-filling curve heuristic, 41, 125
Spencer, J., 24
Spencer, J. H., 112

Square function, 45
Stadje, W., 49
Steele, J. M., 25, 41, 43, 50, 75, 96, 99, 100, 117, 137
Stein, C., 113, 124
Subadditive ergodic theorem, 19
Subadditive Euclidean functionals, 53
    basic theorem, 54
Subadditive sequences, 2, 21, 25
    refinements, 21
Subadditivity
    geometric, 60
    geometric from simple, 64
    simple, 60
Sulanke, R., 116
Syzmansky, T. G., 18
Szegö, G., 25
Szekeres, G., 6

Talagrand, M., 24, 65, 119, 124
Talagrand's convex distance, 120
Talagrand's isoperimetric
    inequality
    proof, 128
Talagrand's isoperimetric
    theorem, 120
Talagrand's isoperimetric
    theory, 119
Terada, R., 41
Traveling-salesman problem (TSP), 27
    Gaussian tail bound, 124
Tusnády, G., 141

van Houweninge, M. V., 93
Veršik, A. M., 10
Vohra, R., 93
Voronoi diagram, 95

Walkup, D. W., 79, 93
Waterman, M. S., 3, 4, 24

Yao, A., 49
Yukich, J. E., 68, 117